T0200659

Active Inference

Active Inference

The Free Energy Principle in Mind, Brain, and Behavior

Thomas Parr, Giovanni Pezzulo, and Karl J. Friston

The MIT Press
Cambridge, Massachusetts
London, England

The MIT Press would like to thank the anonymous peer reviewers who provided comments on drafts of this book. The generous work of academic experts is essential for establishing the authority and quality of our publications. We acknowledge with gratitude the contributions of these otherwise uncredited readers.

This book was set in Stone Serif and Stone Sans by Westchester Publishing Services. Printed and bound in the United States of America.

Library of Congress Cataloging-in-Publication Data is available.

Names: Parr, Thomas, 1993– author. | Pezzulo, Giovanni, author. | Friston, K. J. (Karl J.), author.
Title: Active inference : the free energy principle in mind, brain, and behavior / Thomas Parr, Giovanni Pezzulo, and Karl J. Friston.
Description: Cambridge, Massachusetts : The MIT Press, [2022] | Includes bibliographical references and index.
Identifiers: LCCN 2021023032 | ISBN 9780262045353 (hardcover)
Subjects: LCSH: Perception. | Inference. | Neurobiology. | Human behavior models. | Knowledge, Theory of. | Bayesian statistical decision theory.
Classification: LCC BF311 .P31366 2022 | DDC 153—dc23
LC record available at https://lccn.loc.gov/2021023032

10 9 8 7 6 5

Contents

Preface

Karl Friston

Active Inference is a way of understanding sentient behavior. The very fact that you are reading these lines means that you are engaging in Active Inference—namely, actively sampling the world—in a particular way—because you believe you will learn something. You are palpating this page with your eyes simply because this is the kind of action that will resolve uncertainty about what you will see next and—indeed—what these words convey. In short, Active Inference puts the action into perception, whereby perception is treated as perceptual inference or hypothesis testing. Active Inference goes even further and considers planning as inference—that is, inferring what you would do next to resolve uncertainty about your lived world.

To illustrate the simplicity of Active Inference—and what we are trying to explain—place your fingertips gently on your leg. Keep them there motionless for a second or two. Now, does your leg feel rough or smooth? If you had to move your fingers to evince a feeling of roughness or smoothness, you have discovered a fundament of Active Inference. To feel is to palpate. To see is to look. To hear is to listen. This palpation does not necessarily have to be overt—we can act covertly by directing our attention to this or that. In short, we are not simply trying to make sense of our sensations; we have to actively create our sensorium. In what follows, we will see why this has to be the case and why everything that we perceive, do, or plan is in the compass of one existential imperative—self-evidencing.

Active Inference is not just about reading or epistemic foraging. It is, on one view, something that all creatures and particles do, in virtue of their existence. This might sound like a strong claim; however, it speaks to the fact that Active Inference inherits from a free energy principle that equates existence with self-evidencing and self-evidencing with an enactive sort of inference. However, this book is not concerned with the physics of sentient systems. Its focus is on the implications of this physics for understanding how the brain works.

This understanding is not an easy business, as witnessed by millennia of natural philosophy and centuries of neuroscience. Although one can find the roots of Active Inference in first principle accounts of self-organized behavior (i.e., variational principles akin to Hamilton's principle of stationary action), first principles do not help very much when asking how a particular brain works and how it differs from another brain. For example, committing to the theory of evolution by natural selection does not help in the slightest · when it comes to understanding why I have two eyes or speak French. This book is about using principles to scaffold key questions in neuroscience and artificial intelligence. To do this, we have to move beyond principles and get to grips with the mechanics to which the principles apply.

As such, Active Inference—and its accompanying Bayesian mechanics— is there to frame questions about how we perceive, plan, and act. Crucially, it does not aim to replace other frameworks, such as behavioral psychology, decision theory, and reinforcement learning. Rather, it hopes to embrace all those approaches that have proven so successful within a unified framework. In what follows, we will pay special attention to linking key constructs from psychology, cognitive neuroscience, enactivism, ethology, and so on to the calculus of belief updating in Active Inference—and its associated process theories.

By *process theories*, we refer to theories about how belief updating is realized by neuronal (and other biophysical) processes in the embodied brain and beyond. Work to date in Active Inference offers a fairly straightforward set of computational architectures and simulation tools to both model various aspects of a functioning brain and enable people to test hypotheses about different computational architectures. However, these tools only solve half the problem. At the heart of Active Inference lies a generative model— namely, a probabilistic representation of how unobservable causes in the world out there generate the observable consequences—our sensations. Getting the generative model right—as an apt explanation for the sentient behavior of any experimental subject or creature—is the big challenge.

This book tries to explain how to meet this challenge. The first part sets up the basic ideas and formalisms that are called on in the second part—to illustrate how they can be applied in practice. In short, this book is for people who want to use Active Inference to simulate and model sentient behavior, in the service of either scientific inquiry or, possibly, artificial intelligence. Thus it focuses on those ideas and procedures that are necessary to understand and implement an Active Inference scheme without getting distracted by the physics of sentient systems on the one hand or philosophy on the other.

A Note from Karl Friston

I have a confession to make. I did not write much of this book. Or, more precisely, I was not allowed to. This book's agenda calls for a crisp and clear writing style that is beyond me. Although I was allowed to slip in a few of my favorite words, what follows is a testament to Thomas and Giovanni, their deep understanding of the issues at hand, and, importantly, their theory of mind—in all senses.

Acknowledgments

We gratefully acknowledge invaluable input from our friends and colleagues—in particular, past and present members of the Theoretical Neurobiology group at the Wellcome Centre for Human Neuroimaging, University College London; the Cognition in Action (CONAN) Lab at the Institute of Cognitive Sciences and Technologies, National Research Council of Italy; and numerous international collaborators who have been integral to the development of the ideas presented in this book. This young but growing community has been more than generous in providing both intellectual support and motivation. Furthermore, we gratefully acknowledge Robert Prior and Anne-Marie Bono from MIT Press for kindly accompanying and advising us during the preparation of this book and Jakob Hohwy and other thoughtful reviewers for their guidance. Finally, we thank the funding agencies that provided financial support for our research: KJF was funded by a Wellcome Trust Principal Research Fellowship (Ref: 088130/Z/09/Z); GP was funded by the European Research Council under the Grant Agreement No. 820213 (ThinkAhead) and the European Union's Horizon 2020 Framework Programme for Research and Innovation under the Specific Grant Agreement No. 945539 (Human Brain Project SGA3).

I

1 Overview

Chance favors the prepared mind.

—Louis Pasteur

1.1 Introduction

This chapter introduces the main question that Active Inference seeks to address: How do living organisms persist while engaging in adaptive exchanges with their environment? We discuss the motivation for addressing this question from a normative perspective, which starts from first principles and then unpacks their cognitive and biological implications. Furthermore, this chapter briefly introduces the structure of the book, including its subdivision into two parts: the first of which aims to help readers understand Active Inference, and the second of which aims to help them use it in their own research.

1.2 How Do Living Organisms Persist and Act Adaptively?

Living organisms constantly engage in reciprocal interactions with their environment (including other organisms). They emit *actions* that change the environment and receive *sensory observations* from it, as schematically illustrated in figure 1.1.

Living organisms can only maintain their bodily integrity by exerting adaptive *control* over the action-perception loop. This means acting to solicit sensory observations that either correspond to desired outcomes or goals (e.g., the sensations that accompany secure nutrients and shelter for simple

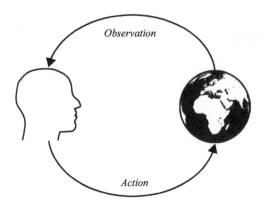

Figure 1.1
An action-perception cycle reciprocally connecting a creature and its environment.
The term *environment* is intentionally generic. In the examples that we discuss, it can
include the physical world, the body, the social environment, and so on.

organisms, or friends and jobs for more complex ones) or help in making
sense of the world (e.g., informing the organism about its surroundings).

Engaging in adaptive action-perception loops with the environment
poses formidable challenges to living organisms. This is largely due to the
recursive nature of the cycle, where each observation, solicited by the previous action, changes how we decide on the next action, to solicit the next
observation. The possibilities for control and adaptation are plentiful, but
very few are useful. Yet during evolution, living organisms have man-
aged to develop adaptive strategies to face the fundamental challenges of
existence. These strategies vary in their level of cognitive sophistication,
with simpler and more rigid solutions in simpler organisms (e.g., following nutrient gradients in bacteria) and more cognitively demanding and
flexible solutions in more advanced organisms (e.g., planning to achieve
distal goals in humans). These strategies also vary for the timescales at
which they are selected and operate—ranging from simple responses to
environmental threats or morphological adaptations that arise at an evo-
lutionarily timescale, to behavioral patterns established during cultural
or developmental learning, up to those requiring cognitive processes that
operate at comparable timescales to action and perception (e.g., attention
and memory).

1.3 Active Inference: Behavior from First Principles

This diversity is a blessing for biology but challenging for formal theories of brain and mind. Broadly, there are two perspectives we could take on this. One perspective is that different biological adaptations, neural processes (e.g., synaptic exchanges and brain networks), and cognitive mechanisms (e.g., perception, attention, social interaction) are highly idiosyncratic and require dedicated explanations. This would lead to proliferation of theories in fields like philosophy, psychology, neuroscience, ethology, biology, artificial intelligence, and robotics, with little hope for their unification. Another perspective is that, despite their diverse manifestations, the central aspects of behavior, cognition, and adaptation in living organisms are amenable to a coherent explanation from first principles.

These two possibilities map to two different research programs and, to some extent, different attitudes toward science: "neats" versus "scruffies" (terms due to Roger Shank). Neats always seek unification beyond the (apparent) heterogeneity of brain and mind phenomena. This usually corresponds to designing top-down, normative[1] models that start from first principles and try to derive as much as possible about brains and minds. Scruffies instead embrace the heterogeneity by focusing on details that demand dedicated explanations. This usually corresponds to designing bottom-up models that start from data and use whatever works to explain complex phenomena, including different explanations for different phenomena.

Is it possible to explain heterogenous biological and cognitive phenomena from first principles, as the neats assume? Is a unified framework to understand brain and mind possible?

This book answers these questions affirmatively and advances Active Inference as a normative approach to understand brain and mind. Our treatment of Active Inference starts from first principles and unpacks their cognitive and biological implications.

1.4 Structure of the Book

The book comprises two parts. These are aimed at readers who want to understand Active Inference (first part) and those who seek to use it for their own research (second part). The first part of the book introduces Active Inference both conceptually and formally, contextualizing it within current theories of cognition. The goal of this first part is to provide a comprehensive, formal, and self-contained introduction to Active Inference: its main constructs and implications for the study of brain and cognition.

The second part of the book illustrates specific examples of computational models that use Active Inference to explain cognitive phenomena, such as perception, attention, memory, and planning. The goal of this second part is to help readers both understand existing computational models using Active Inference and design novel ones. In short, this book divides into theory (part 1) and practice (part 2).

1.4.1 Part 1: Active Inference in Theory

Active Inference is a normative framework to characterize Bayes-optimal[2] behavior and cognition in living organisms. Its normative character is evinced in the idea that all facets of behavior and cognition in living organisms follow a unique imperative: *minimizing the surprise of their sensory observations*. *Surprise* has to be interpreted in a technical sense: it measures how much an agent's current sensory observations differ from its preferred sensory observations—that is, those that preserve its integrity (e.g., for a fish, being in the water). Importantly, minimizing surprise is not something that can be done by passively observing the environment: rather, agents must adaptively *control* their action-perception loops to solicit desired sensory observations. This is the active bit of Active Inference.

Minimizing surprise turns out to be a challenging problem for technical reasons that will become apparent later. Active Inference offers a solution to this problem. It assumes that even if living organisms cannot directly minimize their surprise, they can minimize a proxy—called *(variational) free energy*. This quantity can be minimized through neural computation in response to (and in anticipation of) sensory observations. This emphasis on free energy minimization discloses the relation between Active Inference and the (first) principle that motivates it: the *free energy principle* (Friston 2009).

Free energy minimization seems a very abstract starting point to explain biological phenomena. However, it is possible to derive a number of formal and empirical implications from it and to address a number of central questions in cognitive and neural theory. These include how the variables involved in free energy minimization may be encoded in neuronal populations; how the computations of minimized free energy map to specific cognitive processes, such as perception, action selection, and learning; and what kind of behaviors emerge when an Active Inference agent minimizes its free energy.

As the above list of topics exemplifies, in this book we are mainly concerned with Active Inference and free energy minimization at the level of living organisms—simpler (e.g., bacterial) or more complex (e.g., human)—and their behavioral, cognitive, social, and neural processes. This clarification is necessary to contextualize our treatment of Active Inference within the more general free energy principle (FEP), which discusses free energy minimization across a much wider range of biological phenomena and timescales beyond neural information processing—ranging from evolutionary to cellular and cultural (Friston, Levin et al. 2015; Isomura and Friston 2018; Palacios, Razi et al. 2020; Veissière et al. 2020)—which are beyond the scope of this book.

It is possible to motivate Active Inference by taking one of two roads: a high road and a low road; see figure 1.2. These two roads provide two distinct but highly complementary perspectives on Active Inference:

- The *high road* to Active Inference starts from the question of how living organisms persist and act adaptively in the world and motivates

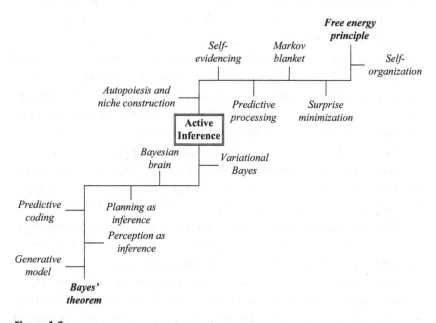

Figure 1.2
Two roads to Active Inference: the high road (starting from top-right) and the low road (starting from bottom-left).

Active Inference as a normative solution to these problems. This high road perspective is useful to understand the normative nature of Active Inference: *what* living organisms must do to face their fundamental existential challenges (minimize their free energy) and *why* (to vicariously minimize the surprise of their sensory observations).

• The *low road* to Active Inference starts from the notion of the Bayesian brain, which casts the brain as an inference engine trying to optimize probabilistic representations of the causes of its sensory input. It then motivates Active Inference as a specific, variational approximation to the (otherwise intractable) inferential problem, which has a degree of biological plausibility. This low road perspective is useful to illustrate *how* Active Inference agents minimize their free energy—therefore illustrating Active Inference not just as a principle but also as a mechanistic explanation (aka *process theory*) of cognitive functions and their neuronal underpinnings.

In chapter 2, we set out the low road perspective on Active Inference. We start from foundational theories that cast perception as a problem of statistical (Bayesian) inference (Helmholtz 1866) and their modern incarnation in the Bayesian brain hypothesis (Doya 2007). We will see that to perform such (perceptual) inference, living organisms must be equipped with—or embody—a *probabilistic generative model* of how their sensory observations are generated, which encodes beliefs (probability distributions) about both observable variables (sensory observations) and nonobservable (hidden) variables. We will extend this inferential view beyond perception to cover problems of action selection, planning, and learning.

In chapter 3, we will illustrate the complementary high road perspective on Active Inference. This chapter introduces the FEP and the imperative for biological organisms to minimize surprise. Further to this, it unpacks how this principle encompasses the dynamics of self-organization and the preservation of a statistical boundary or *Markov blanket* that maintains separation from the environment. This is vital in maintaining the integrity of biological creatures, and it is central to their autopoiesis.

In chapter 4, we will unpack Active Inference more formally. This chapter takes its cue from the discussion of the Bayesian brain in chapter 2 and sets out the mathematical relationship between the self-evidencing dynamics of chapter 3 and variational inference. In addition, this chapter sets out two sorts of generative model used to formulate Active Inference problems. These include the partially observed Markov decision processes used for decision-making and planning and the continuous time dynamical models that interface with sensory receptors and muscles. Finally, we see how free energy minimization for each of these models manifests as dynamic belief updating.

In chapter 5, we will move from formal treatments to biological implications of Active Inference. By starting from the premise that "everything that changes in the brain must minimize free energy" (Friston 2009), we will discuss how the specific quantities involved in the free energy minimization (e.g., prediction, prediction error, and precision signals) manifest in neuronal dynamics. This aids in mapping the abstract computational principles of Active Inference to specific neural computations that can be executed by physiological substrates. This is important in forming hypotheses under this framework and ensures that these are answerable to measured data. In other words, chapter 5 sets out the process theory associated with Active Inference.

Throughout the first part of the book, we will discuss several characteristic aspects of Active Inference. These highlight the ways in which it is different from alternative frameworks that seek to explain biological regulation and cognition—some of which we preview here.

- Under Active Inference, *perception* and *action* are two complementary ways to fulfill the same imperative: minimization of free energy. Perception minimizes free energy (and surprise) by (Bayesian) belief updating or *changing your mind*, thus making your beliefs compatible with sensory observations. Instead, action minimizes free energy (and surprise) by *changing the world* to make it more compatible with your beliefs and goals. This unification of cognitive functions marks a fundamental difference between Active Inference and other approaches that treat action and perception in isolation from one another. Learning is yet another way to minimize free energy. However, it is not fundamentally different from perception; it simply operates at a slower timescale. The complementarity between perception and action will be unpacked in chapter 2.

- In addition to driving action selection in the present to change currently available sensory data, the Active Inference framework accommodates planning—or the selection of the optimal course of action (or policy) in the future. Optimality here is measured in relation to an *expected free energy* and is distinct from the notion of *variational free energy* considered above in the context of action and perception. Indeed, while computing variational free energy depends on present and past observations, computing expected free energy also requires predicted future observations (hence the term *expected*). Interestingly, the expected free energy of a policy comprises two parts. The first quantifies the extent to which the policy is expected to resolve uncertainty (exploration) and the second how consistent the predicted outcomes are with an agent's goals (exploitation). In contrast with other frameworks, policy selection in Active Inference automatically balances exploration and exploitation. The relations between variational and expected free energy will be unpacked in chapter 2.

• Under Active Inference, all cognitive operations are conceptualized as inference over *generative models*—in keeping with the idea that the brain performs probabilistic computations—aka *the Bayesian brain hypothesis*. Yet, the appeal to a specific approximate form of Bayesian inference— that is, a variational scheme that is motivated by first principles—adds specificity to the process theory. Furthermore, Active Inference extends the inferential approach to domains of cognition that are rarely considered and adds some specificity to the kind of models and inferential processes that may be implemented by biological brains. Under some assumptions, the dynamics that emerge from generative models used in Active Inference closely correspond to widespread models in computational neuroscience, such as predictive coding (Rao and Ballard 1999) and the Helmholtz machine (Dayan et al. 1995). The specifics of the variational scheme will be unpacked in chapter 4.

• Under Active Inference, both perception and learning are *active* processes, for two reasons. First, the brain is essentially a *predictive machine*, which constantly predicts incoming stimuli rather than passively waiting for them. This is important as perceptual and learning processes are always contextualized by prior predictions (e.g., expected and unexpected stimuli affect perception and learning in different ways). Second, creatures engaging in Active Inference actively seek out *salient* sensory observations that resolve their uncertainty (e.g., by orienting their sensors or selecting learning episodes that are informative). The active character of perception and learning stands in contrast with most current theories that treat them as largely passive processes; this will be unpacked in chapter 2.

- Action is quintessentially goal directed and purposive. It starts from a desired outcome or goal (analogous to the concept of a set-point in cybernetics), which is encoded as a prior prediction. Planning proceeds by inferring an action sequence that fulfills this prediction (or equivalently, reduces any prediction error between prior prediction and the current state). The goal-directed character of action in Active Inference is in keeping with early cybernetic formulations but is distinct from most current theories that explain behavior in terms of stimulus-response mappings or state-action policies. Stimulus-response or habitual behavior then becomes a special case of a broader family of policies in Active Inference. The goal-directed nature of Active Inference will be unpacked in chapters 2 and 3.

- Various constructs of Active Inference have plausible biological analogues in the brain. This implies that—once one has defined a specific generative model for a problem at hand—one can move from Active Inference as a normative theory to Active Inference as a process theory, which makes specific empirical predictions. For example, perceptual inference and learning correspond to changing synaptic activity and changing synaptic efficacy, respectively. Precision of predictions (in predictive coding) corresponds to the synaptic gain of prediction error units. Precision of policies corresponds to dopaminergic activity. Some of the biological consequences of Active Inference will be unpacked in chapter 5.

1.4.2   Part 2: Active Inference in Practice

While the first part of the book provides readers with the conceptual and formal tools to understand Active Inference, the second part focuses on practical issues. Specifically, we hope to provide readers with the tools to understand existing Active Inference models of cognitive functions (and dysfunctions) and to design novel ones. To this aim, we discuss specific examples of models using Active Inference. Importantly, models of Active Inference can vary along different dimensions (e.g., with discrete or continuous time formulations, flat or hierarchical inference). The second part is structured as follows:

In chapter 6, we introduce a recipe to build Active Inference models. The recipe covers the essential steps to design an effective model, which include the identification of the system of interest, the most appropriate form of the generative model (e.g., to characterize discrete- or continuous-time phenomena), and the specific variables to be included in the model. This chapter therefore offers an introduction to the design principles that underwrite the models discussed in the following chapters.

In chapter 7, we discuss Active Inference models that address problems formulated in discrete time; for example, as hidden Markov models (HMMs) or partially observable Markov decision processes (POMDPs). Our examples include a model of perceptual processing and a model of discrete foraging choices—that is, whether to turn left or right at a decision point to secure a reward. We also introduce topics such as information seeking, learning, and novelty seeking, which can be treated in terms of discrete-time Active Inference.

In chapter 8, we discuss Active Inference models that address problems formulated in continuous time, using stochastic differential equations. These include models of perception (like predictive coding), movement control, and sequential dynamics. Interestingly, it is in the continuous-time formulation that some of the most distinctive predictions of Active Inference appear, such as the idea that movement generation stems from the fulfillment of predictions and that attentional phenomena can be understood in terms of precision control. We also introduce hybrid models of Active Inference that include both discrete- and continuous-time variables. These permit simultaneous assessment of the choice among discrete options (e.g., targets for saccades) and the continuous movements resulting from the choice (e.g., oculomotor movements).

In chapter 9, we illustrate how to use Active Inference models to analyze data from behavioral experiments. We discuss the specific steps that are necessary for model-based data analysis, from the collection of data to the formulation of a model and its inversion to support the analysis of data from single participants or at the group level.

In chapter 10, we discuss the relations between Active Inference and other theories in psychology, neuroscience, AI, and philosophy. We also highlight the most important aspects of Active Inference that distinguish it from the other theories.

In the appendixes, we briefly discuss the mathematical background required to understand the most technical parts of the book, including the notions of Taylor series approximation, variational Laplace, variational calculus, and more. For reference we also present in a concise form the most important equations used in Active Inference.

In sum, the second part of the book illustrates a broad variety of models of biological and cognitive phenomena that can be constructed using Active Inference and a methodology to design novel ones. Apart from the interest of the specific models, we hope that our treatment clarifies the value of using a unified, normative framework to address biological and cognitive phenomena from a coherent perspective. In the end, this is the real appeal of normative frameworks: to provide a unified perspective and a guiding principle to reconcile apparently disconnected phenomena—in this case, phenomena like perception, decision-making, attention, learning, and movement control, each having its separate chapter in any psychology or neuroscience manual.

The models highlighted in the second part have been selected to illustrate specific points as simply as possible. While we cover several models and domains, from discrete-time decisions to continuous-time perception and movement control, we are clearly disregarding many others that are equally interesting. Many other Active Inference models exist in the literature that cover domains as diverse as biological self-organization and the origins of life (Friston 2013), morphogenesis (Friston, Levin et al. 2015), cognitive robotics (Pio-Lopez et al. 2016, Sancaktar et al. 2020), social dynamics and niche construction (Bruineberg, Rietveld et al. 2018), the dynamics of synaptic networks (Palacios, Isomura et al. 2019), learning in biological networks (Friston and Herreros 2016), and psychopathological conditions, such as post-traumatic stress disorder (Linson et al. 2020) and panic disorder (Maisto, Barca et al. 2021). These models vary along many dimensions: some are more directly related to biology whereas others are less so; some are single-agent models whereas others are multi-agent models; some target adaptive inference whereas other target maladaptive inference (e.g., in patient groups), and so on.

This growing literature exemplifies the increasing popularity of Active Inference and the possibility of using it in a very large variety of domains. The aim of this book is to provide our readers with the ability to understand and use Active inference in their own research—possibly, to explore its unforeseen potentialities.

1.5 Summary

This chapter briefly introduces the Active Inference approach to explain biological problems from a normative perspective—and previews some implications of this perspective that will be unpacked in later chapters. Furthermore, this chapter highlights the division of the book into two parts, which aim to help readers understand Active Inference and use it in their own research, respectively. Over the next few chapters, we will develop the low road and high road perspectives outlined herein, before delving into the structure of generative models and the resulting message passing. Together these comprise Active Inference in principle and provide the preliminaries for Active Inference in practice. We hope that these chapters will persuade readers that Active Inference offers not only a unifying principle under which to understand behavior but also a tractable approach to studying action and perception in autonomous systems.

2 The Low Road to Active Inference

My thinking is first and last and always for the sake of my doing.
—William James

2.1 Introduction

This chapter introduces Active Inference by starting from the Helmholtzian—
or perhaps Kantian—view of "perception as unconscious inference" (Helmholtz 1867) and related ideas that have emerged more recently under the
Bayesian brain hypothesis. It explains how Active Inference subsumes and
extends these ideas by treating not just perception but also action, planning,
and learning as problems of (Bayesian) inference and by deriving a principled
(variational) approximation to such otherwise intractable problems.

2.2 Perception as Inference

There is a long tradition of seeing the brain as a "predictive machine,"
or a statistical organ that infers and predicts external states of the world.
This idea dates back to the notion of "perception as unconscious inference" (Helmholtz 1866). More recently, this has been reformulated as the
"Bayesian brain" hypothesis (Doya 2007). From this perspective, perception is not a purely bottom-up transduction of sensory states (e.g., from the
retina) into internal representations of what is out there (e.g., as patterns
of neuronal activity). Rather, it is an inferential process that combines (top-
down) prior information about the most likely causes of sensations with
(bottom-up) sensory stimuli. Inferential processes operate on probabilistic

representations of states of the world and follow Bayes' rule, which prescribes the (optimal) update in the light of sensory evidence. Perception is not a passive outside-in process—in which information is extracted from impressions on our sensory epithelia from "out there." It is a constructive inside-out process—in which sensations are used to confirm or disconfirm hypotheses about how they were *generated* (MacKay 1956, Gregory 1980, Yuille and Kersten 2006, Neisser 2014, A. Clark 2015).

In turn, performing Bayesian inference requires a *generative model*— sometimes referred to as a *forward model*. A generative model is a construct from statistical theory that generates predictions about observations. It may be formulated as the *joint probability $P(y, x)$* of *observations y* and the world's *hidden states x* that generate these observations. The latter are referred to as *hidden* or *latent* states as they cannot be observed directly. This joint probability can be decomposed into two parts. The first is a *prior P(x)*, which denotes the organism's knowledge about the hidden states of the world prior to seeing sensory data. The second is the *likelihood $P(y|x)$*, which denotes the organism's knowledge of how observations are generated from states. Bayes' rule tells us how to combine these two elements, essentially updating a *prior probability P(x)* into a *posterior probability* of hidden states after receiving observations $P(x|y)$. For readers who need a brief refresher on basic probability theory, box 2.1 provides a summary.

Bayesian inference is a broad topic that arises in disciplines like statistics, machine learning, and computational neuroscience. A full treatment of the associated topics is beyond the scope of this book, but there are excellent resources available for those who wish to understand it in depth (Murphy 2012). However, all of this is based on one simple rule. To illustrate this rule, we consider an example of Bayesian perceptual inference (figure 2.1). Imagine a person who has a strong belief that she is confronted with an apple. This belief corresponds to a prior probability, or *prior* for short. This prior comprises the probability attributed to the apple hypothesis and the probability assigned to alternative hypotheses. In this example, our alternative hypothesis is that it is not an apple but a frog. Numerically, the prior probability distribution assigns 0.9 to apple and 0.1 to frog. Note that, as we have assumed that there are only two plausible (mutually exclusive) hypotheses, they must sum to one. The person is also equipped with a *likelihood* model, which assigns a high probability to the fact that frogs jump, whereas apples do not. This likelihood specifies the (probabilistic) mapping from the two hidden states (frog or apple) to the two observations (jumps or does not

Box 2.1
The sum and product rules of probability

Probabilistic reasoning is underwritten by two key rules: the sum and product rules of probability, which are as follows (respectively):

$$\sum_x P(x) = 1$$
$$P(x)P(y|x) = P(x,y)$$

The sum rule says that the probability of all possible events (x) must sum (or integrate) to one. The product rule says that the *joint* probability of two random variables (x and y) may be decomposed into the product of the probability of one variable ($P(x)$) and the *conditional* probability of the second variable given the first ($P(y|x)$). A conditional probability is the probability of one variable (here, y) if we know the value that the other variable (here, x) takes.

We can develop two important results from these simple rules. The first is the operation of *marginalization*. The second is Bayes' rule. Marginalization allows us to obtain a distribution of just one of the two variables from a joint distribution:

$$\underbrace{\sum_x P(x,y) = \sum_x P(y)P(x|y)}_{\text{Product rule}} = \underbrace{P(y)\sum_x P(x|y) = P(y)}_{\text{Sum rule}}$$

The probability of y is referred to as a marginal probability, and we refer to this operation as marginalizing out x. Bayes' rule may be obtained directly from the product rule:

$$\underbrace{P(x)P(y|x) = P(x,y)}_{\text{Product rule}} = \underbrace{P(y)P(x|y)}_{\text{Product rule}}$$

This lets us translate between a prior and conditional distribution (likelihood) and the associated marginal and the other conditional distribution (posterior). Put simply, Bayes' rule just says that the probability of two things is the probability of the first, given the second, times the probability of the second, which is the same as the probability of the second, given the first, times the probability of the first.

jump). Together, the prior and the likelihood form the person's generative model.

Now imagine that the person observes that her apple-frog jumps. Bayes' rule tells us how to form a posterior belief from the prior, taking into account the likelihood of jumping. This rule is expressed as follows:

$$P(x|y) = \frac{P(x)P(y|x)}{P(y)} \tag{2.1}$$

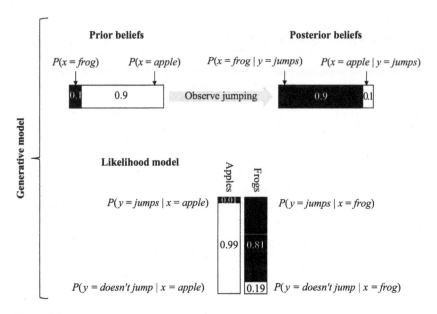

Figure 2.1
A simple example of Bayesian inference. *Upper left:* The organism's prior belief $P(x)$
about the object it will see, before having made any observations, i.e., a categorical
distribution over two possibilities, apple (with probability 0.9) and frog (with prob-
ability 0.1). *Upper right:* The organism's posterior belief $P(x|y)$ after observing that the
object jumps. Posterior beliefs can be computed using Bayes' rule under a likelihood
function $P(y|x)$. This is shown below the prior and posterior, and specifies that if
the object is an apple, there is a very small probability (0.01) that it will jump, while
if it is a frog, the probability that it will jump is much higher (0.81). (The probability
bars in this figure are not exactly to scale.) In this specific case, the update from prior
to posterior is large.

Under the likelihood model in figure 2.1, the posterior probability
assigned to the frog is 0.9, and the probability assigned to the apple is 0.1.
As highlighted in box 2.1, the denominator of equation 2.1 may be computed by marginalizing the numerator. Using our apple-frog example, we
take the opportunity to unpack two different notions of *surprise*—both of
which are important in Active Inference. The first, which we refer to simply
as surprise, is the negative log evidence, where evidence is the marginal
probability of observations. In our example, this is the negative log probability of observing anything jumping under the generative model. Surprise
is a very important quantity from a Bayesian perspective. It is a measure of
how poorly a model fits the data it tries to explain. To put this intuitively,

we can work out the probability of the observed (jumping) behavior under our model. Remember that this assigns a very high prior probability to apples and a low prior probability to frogs. Thus, our marginal probability of jumping is as follows:

$$
\begin{aligned}
P(y = jumps) &= \sum_x P(x, y = jumps) \\
&= \sum_x P(x)P(y = jumps \,|\, x) \\
&= P(x = frog)P(y = jumps \,|\, x = frog) \\
&\quad + P(x = apple)P(y = jumps \,|\, x = apple) \\
&= 0.1 \times 0.81 + 0.9 \times 0.01 \\
&= 0.09
\end{aligned}
\tag{2.2}
$$

This means that, under this model, we would only expect to observe jumping behavior about 9 times out of 100 observations. As such, we should be surprised to observe this if we subscribed to the model in figure 2.1. We can quantify this in terms of surprise ($\mathfrak{S}$). This is given by $\mathfrak{S}(y = jumps) = -\ln P(y = jumps) = -\ln(0.09) = 2.4$ nats.[1] The bigger this number, the worse the model as an apt explanation for the observations at hand. This lets us compare models in relation to data. For example, consider an alternative model, where we have a prior belief that frogs are seen 100 percent of the time. Following the same steps as in equation 2.2, we calculate a surprise of about 0.2 nats. This is a better model of these data, as the observation is much less surprising. The procedure of scoring models on the basis of their evidence (or surprise) is often referred to as Bayesian model comparison. For more complicated models, the form of the surprise may not be so simple. Table 2.1 provides the form of the surprise (omitting constants) for a range of probability distributions—in addition to the categorical probability in our example. Crucially, this lets us talk about surprise for probability distributions whose support[2] differs from the simple example used here. This is important because the way in which sensory data are generated by the world varies with the sort of data. We could be surprised by encountering the face of someone we did not expect to see (categorical distribution), or we could be surprised by it being colder outside than we anticipated (continuous distribution). Table 2.1 may be seen as a portfolio of the probability distributions at our disposal when we come to construct generative models in subsequent chapters. More generally, it makes the point that surprise is a concept that can be evaluated for any given family of probability distributions.

Table 2.1
Probability distributions and surprise[3]

Distribution	Support	Surprise ($\Im$)
Gaussian	$x \in \mathbb{R}$	$\frac{1}{2}(x-\mu) \cdot \Pi(x-\mu)$
Multinomial[1]	$x_i \in (0, \dots, N)$ $i \in \{1, \dots, K\}$ $\sum_i x_i = N$	$-\sum_i x_i \ln d_i$
Dirichlet[2]	$x_i \in (0,1)$ $i \in \{1, \dots, K\}$ $\sum_i x_i = 1$	$\sum_i (1 - \alpha_i) \ln x_i$
Gamma	$x \in (0, \infty)$	$(bx + (1-a)\ln x)$

Notes: 1. Special cases include categorical ($K > 2$, $N = 1$), binomial ($K = 2$, $N > 1$), and Bernoulli ($K = 2$, $N = 1$) distributions. 2. A special case is the beta distribution ($K = 2$).

The second notion of surprise is (slightly confusingly) referred to as *Bayesian surprise*. This is a measure of how much we have to update our beliefs following an observation. In other words, Bayesian surprise quantifies the difference between a prior and a posterior probability. This raises the question of how we quantify the dissimilarity of two probability distributions. One answer, from information theory, is to use a Kullback-Leibler (KL) Divergence. This is defined as the average difference between two log probabilities:

$$D_{KL}\left[Q(x) \| P(x)\right] \triangleq \mathbb{E}_{Q(x)}\left[\ln Q(x) - \ln P(x)\right] \tag{2.3}$$

The $\mathbb{E}$ symbol here indicates an average (or expectation) as outlined in box 2.2. Using the KL-Divergence, we can quantify the Bayesian surprise of our example:

$$
\begin{aligned}
&D_{KL}\left[P(x|y) \| P(x)\right] \\
&= P(x = frog | y = jumps)\left(\ln P(x = frog | y = jumps) - \ln P(x = frog)\right) \\
&\quad + P(x = apple | y = jumps)\left(\ln P(x = apple | y = jumps) - \ln P(x = apple)\right) \\
&= 0.9\left(\ln(0.9) - \ln(0.1)\right) + 0.1\left(\ln(0.1) - \ln(0.9)\right) \\
&\approx 1.8 \text{ nats}
\end{aligned}
\tag{2.4}
$$

This scores the amount of belief updating, as opposed to simply how unlikely the observation was. To highlight the distinction between surprise and Bayesian surprise, consider what happens if we commit to a prior belief that we will always see apples. The Bayesian surprise will be zero, as the prior

Box 2.2

Expectations

It is useful to refer to the *expectation* of a random variable x, usually denoted $\mathbb{E}[x]$. This is the weighted average of all the values that the variable can assume, weighted by their probability. For discrete random variables (that can only take a countable number of possible values), this is given by a weighted sum:

$$\mathbb{E}[x] = \sum_x xP(x)$$

For example, for a discrete (numerical) variable that can only assume two values (1 and 2) with equal probability of $\frac{1}{2}$, this is $\mathbb{E}[x] = 1 \cdot \frac{1}{2} + 2 \cdot \frac{1}{2} = \frac{3}{2}$.

For continuous random variables (that can take infinitely many values), sums are replaced by integrals. Expectations can also be applied to functions of random variables, as opposed to the variables directly. For example, if we have a function $f(x)$, where x has some continuous distribution, the expectation is defined as follows:

$$\mathbb{E}[f(x)] = \int f(x)p(x)dx$$

We will use this notation throughout this book, where the function $f(x)$ will often be a log probability, or log probability ratio.

---

is so confident that we do not update it at all following our observations. However, the surprise is very large (4.6 nats) as it is highly unlikely that an apple will jump.

Note that while we illustrated Bayesian inference on the basis of a very simple generative model, it applies to generative models of any complexity. In chapter 4 we will highlight two forms of generative model that underwrite most applications in Active Inference.

2.3 Biological Inference and Optimality

There are two important points that connect the above inferential scheme to biological and psychological theories of perception. First, the inferential procedure discussed requires the interplay of top-down processes that encode predictions (from the prior) and bottom-up processes that encode sensory observations (as mediated by the likelihood). This interplay of top-down and bottom-up processes distinguishes the inferential view from alternative approaches that only consider bottom-up processes. Furthermore, it is central in modern biological treatments of perception, such as *predictive coding* (discussed in chapter 4), which is a specific algorithmic (or process-level) implementation of the more general (Bayesian) inference scheme discussed here.

Second, Bayesian inference is *optimal*. Optimality is defined in relation to a cost function that is optimized (i.e., minimized), which, for Bayesian inference, is known as variational free energy—closely related to surprise. We return to this in section 2.5. By explicitly considering the full distribution over hidden states, it naturally handles uncertainty, hence avoiding the limitations of alternative approaches that only consider point estimates of hidden states (e.g., the mean value of x). One such alternative would be maximum likelihood estimation, which simply selects the hidden state most likely to have generated the data at hand. The problem with this is that such estimates ignore both the prior plausibility of the hidden state and the uncertainty surrounding the estimation. Bayesian inference does not suffer these limitations. However, despite the use of surprise in objectively assessing whether the model is fit for purpose, it is important to appreciate that inference itself is *subjective*. The results of inference are not necessarily accurate in any objective sense (i.e., the organism's belief may not actually correspond to reality) for at least two important reasons. First, biological creatures operate on the basis of limited computational and energetic resources, which render exact Bayesian inference intractable.[4] This requires approximations that preclude guarantees of exact Bayesian optimality. These approximations include the notion of a variational posterior—based on something called a mean field approximation—which is central to chapter 4.

The second reason optimality may be thought of as subjective is that organisms operate on the basis of a subject's *generative model* of how their observations are generated, which may or may not correspond to the real *generative process* that generates their observations. This is not to say that the *generative model* should correspond to the *generative process*. In fact, there may be models that afford better (e.g., simpler) explanations of the data at hand than the processes that actually generated them—as quantified by their relative surprise. A nice example of this is illusions, for which someone finds a simpler explanation for their visual input in relation to how the visual stimuli have been carefully engineered by a mischievous psychophysicist.

The generative model itself may be optimized as new experience is acquired. This may or may not converge to the generative process. Figure 2.2 illustrates

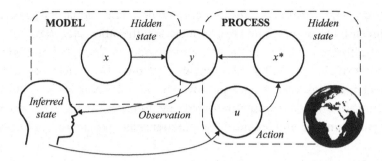

Figure 2.2
Generative process and generative model. Both represent ways in which sensory data
(*y*) could be generated given hidden states (*x*) and are represented through arrows
from *x* to *y* to indicate causality. The difference is that the process is the true causal
structure by which data are generated, while the model is a construct used to draw
inferences about the causes of data (i.e., use observations to derive inferred states).
The hidden states of the generative model and the generative process are not the
same. The organism's model includes a range of hypotheses (*x*) about the hidden
state, which do not necessarily include the true value of the hidden state x^* of the
generative process. In other words, the models we use to explain our sensorium may
include hidden states that do not exist in the outside world, and vice versa. Action
(*u*) is generated on the basis of the inferences made under a generative model. Action is
shown here as part of the generative process, making changes to the world, despite
being selected from the inferences drawn under the model.

this point and the difference between true environmental contingencies, or
the generative process, which is inaccessible to the organism and the organism's generative model of the world. In this particular example, the generative
process is in a true state x^* that is inaccessible to the organism. However, the
organism and world are reciprocally coupled, and x^* generates an observation
y, which the organism senses. The organism can use this observation *y* and
Bayes' rule to infer the (posterior probability of) some explanatory variable
or hidden state in the generative model. In the figure, we refer to both x^*
and *x* as *hidden states*, emphasizing that neither is observable. However, they
are subtly different: the former is part of the organism's generative model,
whereas the latter is part of the generative process and inaccessible to the
organism. Furthermore, x^* and *x* do not necessarily live in the same space. It
might be that the hidden states in the external world take on values that lie
outside the space of explanations available to the brain. Conversely, it might
be that the brain's explanations include variables that do not exist in the

outside world. For example, the former could be 5-dimensional and the latter 2-dimensional, or one could be continuous and the other categorical.

The distinction between the generative model and process is important to contextualize psychological claims about optimality of inference—to the extent that these claims are valid—which, on a Bayesian view, is always contingent on the organism's resources. By resources, we mean its specific generative model, and bounded computational and mnemonic resources.

2.4   Action as Inference

The discussion to this point is common to all Bayesian brain theories. However, we now introduce the simple but fundamental advance offered by Active Inference. This starts from the same inferential perspective discussed above but extends it to consider action as inference. This idea stems from the concept that Bayesian inference minimizes surprise (or, equivalently, maximizes Bayesian model evidence). So far, we have considered what happens when we compute surprise by performing inference—and select among models on the basis of their capacity to minimize surprise. However, surprise does not only depend on the model. It also depends on the data. By acting on the world to change the way in which data are generated, we can ensure a model is fit for purpose by choosing those data that are least surprising under our model.

Equipped with a mechanism to produce actions, an organism can engage in reciprocal exchanges with its environment; see figure 2.2. In animals, this mechanism takes the form of a motor reflex loop. Essentially, for each action-perception cycle, the environment sends an observation to the organism. The organism uses (an approximation to) Bayesian inference to infer its most likely hidden states. It then generates an action and sends it to the environment in an attempt to make the environment less surprising. The environment executes the action, generates a new observation, and sends it to the organism. Then, a new cycle starts. The sequential description here is written for didactic purposes; it is important to realize that these are not really discrete steps but are continuous dynamical processes.

Active Inference goes beyond the recognition that perception and action have the same (inferential) nature. It also assumes that both *perception and action cooperate to realize a single objective*—or optimize just one function—rather than having two distinct objectives, as more commonly assumed. In the Active Inference literature, this common objective has been described in various (informal and formal) ways, including the minimization of surprise, entropy, uncertainty, prediction error, or (variational) free energy. These terms are related to one another but sometimes their relations are not immediately clear, causing some confusion. Furthermore, these terms are used in different contexts; for example, *prediction error* minimization is used in biological contexts where the objective is explaining brain signals, while variational free energy minimization is used in machine learning.

In the next two sections, we will clarify that the single quantity that Active Inference agents minimize through perception and action is *variational free energy*. However, under some conditions, one can reduce variational free energy to other notions, such as the discrepancy between the generative model and the world, or the difference between what one expects and what one observes (i.e., a prediction error). We will introduce variational free energy formally in section 2.6. For simplicity section 2.5 focuses on the ways in which perception and action minimize the discrepancy between the generative model and the world.

2.5 Minimizing the Discrepancy between Model and World

Having established perception and action in terms of Bayesian inference, we now turn to the question of what the objective of inference is. In other words, what is being optimized by inference? In cognitive science, it is common to assume that different cognitive functions like perception and action optimize different objectives. For example, we could assume perception maximizes the accuracy of reconstruction while action selection maximizes utility. Instead, a fundamental insight of Active Inference is that both perception and action serve the very same objective. As a first approximation, this common objective of *perception and action* can be formulated as a minimization of the *discrepancy between the model and the world*. Sometimes this is operationalized in terms of prediction error.

To understand how perception and action reduce the discrepancy between the model and the world, consider again the example of a person who expects to see an apple (figure 2.3). She generates a top-down visual prediction (e.g., about seeing something red and not jumping). This visual prediction is compared with a sensation (e.g., something jumping)—and this comparison results in a discrepancy.

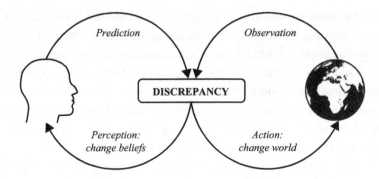

Figure 2.3
Both perception and action minimize discrepancy between model and world.

The person can resolve this discrepancy in two ways. First, she could change her mind about what she is seeing (i.e., a frog) to fit the world, hence resolving the discrepancy. This corresponds to *perception*. Second, she could foveate the nearest apple tree and see something that looks very much like an apple. This also resolves the initial discrepancy, but in a different way. This entails changing the world—including her direction of gaze—and subsequent sensations to fit what is in her mind, not changing her mind to fit the world. This is the other direction of fit. This is *action*.

While changing the direction of one's gaze seems less compelling than changing one's mind in the world of apples and frogs, let us consider another case: a person who expects his body temperature to be in a certain range who senses a high temperature via central thermoreceptors. This is surprising and presents a significant discrepancy to resolve. As in the former example, he has two ways to minimize this discrepancy, corresponding to perception (changing mind) and action (changing the world), respectively. In this case, simply changing one's mind does not seem very adaptive, but acting to lower the body temperature (e.g., by opening the window) is.

This speaks to the fact that in Active Inference, the notion of marginal probabilities or *surprise* (e.g., about body temperature) has a meaning that goes beyond standard Bayesian treatments to absorb notions like homeostatic and allostatic set-points. Technically, Active Inference agents come equipped with models that assign high marginal probabilities to the states they prefer to visit or the observations they prefer to obtain. For a fish, this means a high marginal likelihood for being in water. This implies that organisms implicitly expect the observations they sample to be within their comfort zone (e.g., physiological bounds).

In sum, we have discussed how, at any point in time, we can minimize the discrepancy between our model and our world through perception and action. Whether we adjust our beliefs or our data depends on the confidence with which we hold those beliefs. In our example of the apple, the belief is held with sufficient uncertainty that this will be updated as opposed to acted on. In contrast, in the temperature example, we are considerably more confident about our core temperature because it underwrites our existence. This confidence means we update our world to comply with our beliefs. Yet, in Active Inference, perception and action act more cooperatively than suggested by this treatment. To understand why this is the case, the next section moves from the restricted notion of *discrepancy* (or *prediction error*) to the more general notion of *variational free energy*—which is the quantity that Active Inference actually minimizes and which subsumes prediction error as a special case.

2.6 Minimizing Variational Free Energy

So far, we have discussed perception and action within a Bayesian scheme that aims to minimize surprise. Yet, exact Bayesian inference supporting perception and action is computationally intractable in most cases, because two quantities—the *model evidence* ($P(y)$) and the *posterior probability* ($P(x|y)$)—cannot be computed for two possible reasons. The first is that for complex models, there may be many types of hidden states that all need marginalizing out, making the problem computationally intractable. The second is that the marginalization operation might require analytically intractable integrals. Active Inference appeals to a *variational* approximation of Bayesian inference that is tractable.

The formalism of variational inference will be unpacked in chapter 4. Here, it suffices to say that performing variational Bayesian inference implies substituting the two intractable quantities—*posterior probability* and (log) *model evidence*—with two quantities that approximate them but can be computed efficiently—namely, an *approximate posterior Q* and a *variational free energy F*, respectively. The approximate posterior is sometimes called a variational or recognition distribution. Negative variational free energy is also known as an evidence lower bound (ELBO), especially in machine learning.

Most importantly, the problem of Bayesian inference now becomes a problem of optimization: the minimization of *variational free energy F*. Variational free energy is a quantity with roots in statistical physics that plays a fundamental role in Active Inference. In equation 2.5, it is denoted as $F[Q, y]$, as it is a functional (function of a function) of the approximate posterior Q and a function of data y:

$$
\begin{aligned}
F[Q, y] &= \underbrace{-\mathbb{E}_{Q(x)}[\ln P(y, x)]}_{Energy} - \underbrace{H[Q(x)]}_{Entropy} \\
&= \underbrace{D_{KL}[Q(x) \,\|\, P(x)]}_{Complexity} - \underbrace{\mathbb{E}_{Q(x)}[\ln P(y\,|\,x)]}_{Accuracy} \\
&= \underbrace{D_{KL}[Q(x) \,\|\, P(x\,|\,y)]}_{Divergence} - \underbrace{\ln P(y)}_{Evidence}
\end{aligned}
\tag{2.5}
$$

Variational free energy may seem prima facie an abstract concept, but its nature and the role it plays in Active Inference become apparent when decomposed into quantities that are more intuitive and familiar in cognitive science. Each of these perspectives on variational free energy offers useful intuitions about what free energy minimization means. We briefly sketch these intuitions here, as they will become important when we discuss examples in the second part of the book.

The first line of equation 2.5 shows that minimizing with respect to Q requires consistency with the generative model (*energy*) while also maintaining a high posterior *entropy*.[5] The latter means that, in the absence of data or precise prior beliefs (which only influence the *energy* term), we should adopt maximally uncertain beliefs about the hidden states of the world, in accord with Jaynes's *maximum entropy* principle (Jaynes 1957). Put simply, we should be uncertain (adopt a high entropy belief) when we have no information. The term *energy* inherits from statistical physics. Specifically, under a Boltzmann distribution, the average log probability of a system adopting some configuration is inversely proportional to the energy associated with that configuration—that is, the energy required to move the system into this configuration from a baseline configuration.

The second line emphasizes the interpretation of free energy minimization as finding the best explanation for sensory data, which must be the simplest (minimally *complex*[6]) explanation that is able to *accurately*[7] account for the data (cf. Occam's razor). The complexity-accuracy trade-off recurs across several domains, normally in the context of model comparison for data analysis. In statistics, other approximations to model evidence are sometimes used, such as the Bayesian information criterion or Akaike information criterion. The complexity-accuracy trade-off will become important when we describe how to use free energy for model comparison during model-based data analysis—and in the context of structure learning and model reduction. Inferring explanations that have minimal complexity is also important from a cognitive perspective. This is because one can assume that updating what one knows (the prior) to accommodate the data entails a cognitive cost (Ortega and Braun 2013, Zénon et al. 2019); hence, an explanation that diverges minimally from the prior is preferable.

On this view, the complexity cost is just Bayesian surprise. In other words, the degree to which "I change my mind" is quantified by the divergence between the prior and the posterior. This means every accurate explanation for my sensations incurs a complexity cost, and this cost scores the degree of *Bayesian belief updating*. Variational free energy, then, scores the difference between accuracy and complexity.

The final line expresses the free energy as a bound on negative log evidence (see figure 2.4). As the left part of the figure illustrates, the free energy is an upper bound on negative log evidence, where the bound is the divergence between Q and the posterior probability that would have been obtained were it possible to perform exact (as opposed to variational) inference. The right part of the figure shows that as the divergence decreases, the

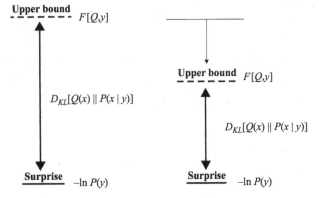

Figure 2.4
Variational free energy as an upper bound on negative log evidence.

free energy approaches the negative log evidence (surprise)—and becomes equal to surprise, if the approximate posterior Q matches the exact posterior $P(x|y)$. This offers a formal motivation for perceptual inference as one way to lower free energy by optimizing our approximate posterior Q as much as possible.

The final line of equation 2.5 shows that perceptual inference is not the only way to minimize free energy. We could also change the log evidence term through acting to change sensory data. This decomposition is interesting from a cognitive perspective, since *minimizing divergence* and *maximizing evidence* map to the two complementary sub-objectives of perception and action, respectively; see figure 2.5. Note that the above expressions all become ways of characterizing the negative log evidence if we replace Q with $P(x|y)$, generalizing to the case of exact inference.

In sum, Active Inference amounts to minimizing variational free energy by perception and action. This minimization permits an organism to fit its generative model to the observations it samples. This fit is a measure of both perceptual adequacy (as expressed by the divergence term) and active control over external states—in the sense that it permits the organism to maintain itself in a suitable set of preferred states, as defined by the generative model. Another way of phrasing this is to appeal to the divergence versus

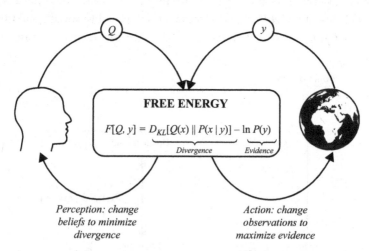

Figure 2.5

Complementary roles of perception and action in the minimization of variational free energy.

evidence decomposition of free energy. Equating the negative log evidence with surprise, and noting that the smallest possible divergence is zero, we see that free energy is an upper bound on surprise. This means it can only be greater than or equal to surprise. When the organism minimizes its divergence (through perception), then free energy becomes an approximation to surprise. When an organism additionally changes the observations it gathers (by acting) to render them more similar to prior predictions, it minimizes surprise.

Variational free energy has a *retrospective* aspect, as it is a function of past and present, but not future, observations. Although it facilitates inferences about the future based on past data, it does not directly facilitate *prospective* forms of inference based on anticipated future data. This is important in planning and decision-making. Here, we infer the best actions or action sequences (policies) on the basis of the future observations they are expected to bring about. Doing this requires that we supplement our generative models with the notion of *expected free energy*.

2.7   Expected Free Energy and Planning as Inference

Expected free energy extends Active Inference to include a quintessentially prospective form of cognition: planning. Planning a sequence of actions, such as the series of moves required to escape from a maze, requires considering future observations that one expects to gather. For example, the consequences of possible courses of action include seeing a dead end after turning right or seeing the exit after a sequence of three left turns. Each possible sequence of actions is termed a *policy*. This highlights an important distinction made in Active Inference between an action and a policy. The former is something that directly influences the outside world, while the latter is a hypothesis about a way of behaving. The implication is that Active Inference treats planning and decision-making as a process of inferring what to do. This brings planning firmly into the realm of Bayesian inference and means we must specify priors and likelihoods as before (section 2.1). However, in place of frogs and apples, the alternatives are behavioral policies (Is it more probable that I look toward the pond or the tree?). In this section, we first briefly deal with the likelihood—that is, the consequences of pursuing a policy—and then turn to the prior. This is where expected free energy comes in.

Policy-dependent outcomes are not immediately available (they are in the future), but they can be predicted by chaining together two components of the generative model. The first is our beliefs about how hidden states change as a function of policies. We will get into the details of this in chapter 4. For now, we use the notation $\tilde{x}$ to denote a sequence or trajectory of hidden states over time, and we condition trajectories on the policies (π) a creature pursues. This means the dynamical part of our model is given by $P(\tilde{x} \mid \pi)$. Drawing from our earlier frog-apple example, the policy may be the decision to go to a pond or to an orchard, which changes the probability of encountering frogs versus apples.

The second component of the model is the usual *likelihood* distribution. This describes which observations to expect in every possible state (e.g., jumping or not, conditioned on frog or apple). By combining these two components, an organism can engage its generative model vicariously to run "what if" or *counterfactual* simulations of the consequences of its possible actions or policies—for example, "What would happen if I go to the pond?" Marginalizing over states, this gives us the marginal likelihood or evidence for a policy ($P(\tilde{y} \mid \pi)$), or a free energy approximation to this quantity. In other words, knowing how policies influence state transitions lets us compute the likelihood of a sequence of observations under that policy. As we saw in equation 2.1, we need to combine this likelihood with a prior probability to calculate the posterior probability of pursuing a policy.

Active Inference decomposes this planning problem into two successive operations. The first is to compute a score for each policy. The second is to form posterior beliefs about which to pursue. The former defines the prior belief about the policies to pursue, where the best policies have high probability and the worst policies have low probability. Under Active Inference, the goodness of a policy is scored by the associated negative expected free energy—just as the goodness of a model fit is scored by the negative free energy of that model. The *expected free energy* (G) of policy is different from the variational free energy (F), since calculating the former requires consideration of future, policy-dependent observations. In contrast, the latter only considers present and past observations. Calculating expected free energy therefore engages the generative model to predict future observations that would stem from each policy—if it were to be executed—up to some planning horizon. Furthermore, because a policy unfolds over multiple time steps, the final measure of expected free energy for each policy has to integrate over all future time steps of that policy.

The expected free energy of each policy can be converted in a quality score (by taking its negative) and is directly available as a prior by agents engaging in Active Inference. This is because—consistent with the notion of potential energy in physics—expected free energy is expressed in the space of log probabilities. Converting it into a belief (or probability distribution) over policies is then a matter of exponentiating (to undo the log) and normalizing (to ensure consistency with the sum rule in box 2.1). Policies that are associated with a lower expected free energy are assigned higher probability and become the policies that the organism expects to pursue.

Ultimately, inferring that we are pursuing a particular policy has consequences for the sensory data we predict. For example, a policy that includes flexing my elbow entails predictions about the proprioceptive input from the biceps and triceps muscles. This provides the link between planning and action, as the predictions associated with a plan translate into action that resolves discrepancies with measured proprioceptive data (see section 2.3).

2.8   What Is Expected Free Energy?

So far, we have assumed that during planning, the organism scores its policies according to their expected free energy. However, we have sidestepped what expected free energy actually is. Like *variational free energy*, the expected free energy can be decomposed in several, mathematically equivalent ways. Each of these provides an alternative perspective on this quantity.

$$
\begin{aligned}
G(\pi) = &-\underbrace{\mathbb{E}_{Q(\tilde{x},\tilde{y}|\pi)}[D_{KL}[Q(\tilde{x}|\tilde{y},\pi)\,\|\,Q(\tilde{x}|\pi)]]}_{\text{Information gain}} - \underbrace{\mathbb{E}_{Q(\tilde{y}|\pi)}[\ln P(\tilde{y}|C)]}_{\text{Pragmatic value}} \\
= &\underbrace{\mathbb{E}_{Q(\tilde{x}|\pi)}[H[P(\tilde{y}|\tilde{x})]]}_{\text{Expected ambiguity}} + \underbrace{D_{KL}[Q(\tilde{y}|\pi)\,\|\,P(\tilde{y}|C)]}_{\text{Risk (outcomes)}} \\
\leq &\underbrace{\mathbb{E}_{Q(\tilde{x}|\pi)}[H[P(\tilde{y}|\tilde{x})]]}_{\text{Expected ambiguity}} + \underbrace{D_{KL}[Q(\tilde{x}|\pi)\,\|\,P(\tilde{x}|C)]}_{\text{Risk (states)}} \\
= &-\underbrace{\mathbb{E}_{Q(\tilde{x},\tilde{y}|\pi)}[\ln P(\tilde{y},\tilde{x}|C)]}_{\text{Expected energy}} - \underbrace{H[Q(\tilde{x}|\pi)]}_{\text{Entropy}}
\end{aligned}
\tag{2.6}
$$

$$Q(\tilde{x},\tilde{y}|\pi) \triangleq Q(\tilde{x}|\pi)P(\tilde{y}|\tilde{x})$$

The first of these is perhaps the most useful, intuitively, as it expresses the value of seeking new information (i.e., exploration) in exactly the same units (nats) as the value of seeking preferred observations (i.e., exploitation), dissolving the classic exploit-explore dilemma in behavioral psychology. By minimizing expected free energy, the relative balance between these terms determines whether behavior is predominantly explorative or exploitative. Note that pragmatic value emerges as a prior belief about observations, where the C-parameter includes preferences. The (potentially unintuitive) link between prior beliefs and preferences is unpacked in chapter 7; for now, we note that this term can be treated as an expected utility or value, under the assumption that valuable outcomes are the kinds of outcomes that characterize each agent (e.g., a body temperature of 37 °C).

The information gain term inherits from the divergence we considered in section 2.5, which ensures that free energy is an upper bound on surprise. However, there is a twist: instead of minimizing the divergence, we want to select policies that maximize the expected divergence—hence, information gain. This switch is due to the fact that we are now taking an average of the log probabilities over outcomes that have yet to be observed. This is a subtle point that can be understood in terms of outcomes switching their roles. When evaluating the free energy of outcomes, the outcomes are the consequences. However, when evaluating the expected free energy, the outcomes play the role of causes in the sense they are variables that are hidden in the future but explain decisions in the present.

The ensuing information gain penalizes observations for which there is a many-to-one mapping from states to observations—in the sense that one can obtain the same observations in different states—as this precludes precise belief updating. In artificial intelligence and robotics, states that bring the same observation (e.g., two T-junctions of a maze that look identical) are sometimes called *aliased* and are generally hard to deal with using simple methods (i.e., stimulus-response, with no inference or memory). The problem is that we cannot know which state we occupy from current observations alone. Active Inference avoids getting into such situations in the first place, given their low potential for information gain.

A simple example may help unpack the distinction between information gain (or epistemic value) and pragmatic value and highlight why, in most realistic situations, pragmatic and epistemic values need to be pursued in tandem. Imagine a person who wants an espresso and knows that there are two good cafes in the town: one that opens only from Monday to Friday and another that opens only during the weekend. If he does not know what day of the week it is, he has to first select an action that has epistemic value and resolves his uncertainty (i.e., an *epistemic action* to look at the calendar)—and only after that select an action that carries pragmatic value and brings the reward (i.e., a *pragmatic action* to go to the correct cafe). This scenario illustrates the fact that in most uncertain situations, one must first perform epistemic actions to resolve uncertainty before confidently selecting a pragmatic action. Policy selection methods that fail to consider the epistemic affordance of choices can only select policies by using random number generators—and will often miss out on their espresso. Therefore, schemes that consider only pragmatic value are generally restricted to situations with no epistemic uncertainty, such as in the case of a person who already knows the day of the week and hence can head directly to the correct cafe.

The second decomposition in equation 2.6 is in terms of *risk* and expected *ambiguity*. These terms are the analogues of *complexity* and *inaccuracy*: risk is the expected complexity, and ambiguity is the expected inaccuracy. Risk, a common notion in economics, corresponds to the fact that there can be a one-to-many mapping between policies and their consequences—in the sense that one can obtain several different outcomes (by chance) under the same policy. One example is a gambling scenario with stochastic rewards (e.g., a one-armed bandit, aka a slot machine), wherein one could know the reward distribution—say, that one will obtain reward 10 percent of the time. This is called a risky situation in economics because, after the same move (pulling a lever), one could obtain two different observations (reward or no reward). This means one has to choose policies or plans that accommodate uncertainty. In risk-sensitive schemes—like active inference—the game is to choose policies whose probabilistic outcomes match, in the sense of a KL-Divergence, one's prior preferences. In short, minimizing complexity cost becomes minimizing risk when both are measures of departure from prior beliefs.

Similarly, *ambiguity* corresponds to the expected inaccuracy due to an ambiguous mapping between states and outcomes. A mapping is ambiguous if the distribution of outcomes anticipated is highly dispersed (or entropic) even if we know the states generating them with complete confidence. For instance, the probability of heads or tails in a coin flip, conditioned on whether it is sunny or raining, will be maximally ambiguous as there is no relationship between the weather and the 50-50 chance of heads or tails. As such, it would not be possible to gain information about the weather by observing tails. Note that most situations are endowed with both risk and ambiguity—which implies a many-to-one mapping between states and outcomes and between policies and outcomes. Recall that outcomes (observations) are the only sort of variable that can be observed. Active Inference deals automatically with these situations, because expected free energy comprises both risk and ambiguity terms.

The third line of equation 2.6 highlights an alternative formulation of the expected free energy by reexpressing risk as a divergence between beliefs about states and preferences defined in terms of states. An appealing feature of this form is that it may be rearranged into an expected energy and entropy in analogy with variational free energy (equation 2.5). While this relationship is attractive, a downside of this formulation is that it assumes the state-space is known a priori such that prior preferences may be associated with states. In most settings, this is not a problem, and the choice between defining preferences in terms of states or outcomes has little practical relevance. However, common practice is to specify preferences in terms of outcomes—allowing the state-space itself to be learned while preserving extrinsic motivation.

In summary, expected free energy can be decomposed in terms of risk and ambiguity and in terms of pragmatic and epistemic values. These decompositions are interesting as they permit a formal understanding of the wide variety of situations that Active Inference deals with. Furthermore, they facilitate an appreciation of how Active Inference subsumes several decision schemes—which may be obtained by ignoring one or more components of expected free energy (figure 2.6). If one removes prior preferences, the pragmatic value becomes irrelevant, and all action is motivated by epistemic affordances—hence such schemes can only handle the resolution of uncertainty. Once prior preferences are removed, the (negative) expected free energy is variously known as expected *Bayesian surprise* (in the context of attentional exploration) or *intrinsic motivation* (in the context of autonomous learning). If one removes ambiguity, the resulting scheme corresponds to *risk-sensitive* or *KL control* in control theory. Finally, if one removes both ambiguity and prior preferences, the only remaining imperative is to maximize the entropy of observations (or states, if using the formulation in the third line of equation 2.6). This may be interpreted as uncertainty sampling (or keeping one's options open). Active Inference evinces the formal relations between these schemes and the (limited) situations in which they apply.

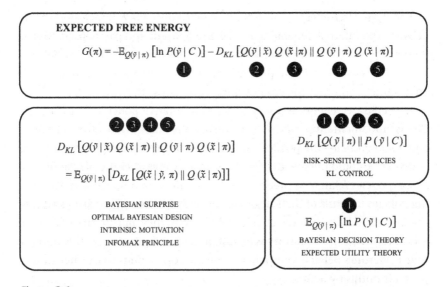

Figure 2.6
Various schemes that can be derived by removing terms from the free energy equation. The upper panel shows the terms contributing to the expected free energy. The lower panels show the schemes that result from removing prior preferences (1), ambiguity (2), or everything except for the prior preferences. Each of these quantities appears in several different fields under a variety of names, but all can be seen as components of the same expected free energy.

Although we have carefully decomposed expected free energy in a way that different people might read this functional, there is no right or wrong way of carving it up. We will see in the second half of this book why autonomous systems of a certain kind must, in virtue of existing, choose actions that look as if they are minimizing expected free energy. This perspective means there is no privileged role for epistemic (explorative) versus pragmatic (exploitative) imperatives—or for risk versus ambiguity. These (possibly false) dichotomies are just two sides of the same existential coin.

2.9 At the End of the Low Road

Having introduced the two distinct notions of *variational free energy* and *expected free energy*, we are now in a position to consider what they achieve together. This represents an endpoint to the low road into Active Inference, starting from the notion of unconscious inference, via the Bayesian brain, the duality of perception and action, and finally planning as inference.

Variational free energy is at the core of Active Inference. It measures the fit between the internal generative model and (current and past) observations. By minimizing variational free energy, creatures maximize their model evidence. This ensures that the generative model becomes a good model of the environment and that the environment complies with the model.

Expected free energy is a way to score alternative policies for planning. This is fundamentally *prospective*—it considers possible future observations— and *counterfactual*—the possible future observations are conditioned on the policies one could pursue. Expected free energy measures the plausibility of action policies relative to preferred (future) states and observations. By scoring policies in terms of their negative expected free energy, creatures engaging in Active Inference effectively believe that they pursue the course of action for which this quantity is lowest. In psychological terms, this implies that a creature's belief about policies directly corresponds to its intention— which it fulfills by acting.

From a conceptual perspective, we can associate minimization of variational free energy and expected free energy with two inferential loops, one nested within the other. Variational free energy minimization is the key (outside) loop of Active Inference, which is sufficient to optimize perception and beliefs about policies. An Active Inference agent can also be endowed with a generative model of the consequences of its action that entails an evaluation of expected free energy (the inside loop). This ability to plan into the future supports prospective forms of action selection by furnishing probability values for policies (Friston, Samothrakis, and Montague 2012; Pezzulo 2012).

2.10 Summary

Active Inference is a theory of how living artifacts underwrite their existence by minimizing surprise—or a tractable proxy to surprise, *variational free energy*—via perception and action. In this chapter, we have sought to motivate this idea starting from a Bayesian treatment of perception as inference and extending this to the domain of action. Bayesian inference rests on a generative model of how sensory observations are generated, which encodes (probabilistically) the organism's implicit knowledge of the world—formalized as prior beliefs and the expected outcomes under alternative states and policies.

The specific take of Active Inference forces us to revisit the usual semantics of a prior in Bayesian inference. Expected states are preferred and include the organism's conditions for survival (e.g., niche-specific goal states), whereas their opposite—surprising states—are dis-preferred. In this way, by fulfilling their expectations, Active Inference agents ensure their own survival. Given the important links between the notion of priors and the conditions that undergird an organism's existence, we can also say that in Active Inference, the identity of an agent is isomorphic with its priors. This terminology will become more familiar later in the book.

Note that in this view, *surprise* (or sometimes *surprisal*) is a formal construct of information theory and not necessarily equivalent to a (folk) psychological construct. Roughly, the more the organism's state differs from the prior (which encodes the preferred states), the more it is surprising— hence Active Inference amounts to the idea that an organism (or its brain) has to actively minimize its surprise to stay alive. Under certain conditions, surprise minimization can be construed as the reduction of the discrepancy between the model and the world. More generally, the quantity that is actually minimized in Active Inference is *variational free energy*. Variational free energy is an (upper-bound) approximation to surprise and can be minimized efficiently using chemical or neuronal message passing and information that is available to the organism's generative model.

Importantly, both perception and action minimize variational free energy in complementary ways: by refining their (posterior belief) estimate and by performing actions that selectively sample what is expected. Furthermore, Active Inference also minimizes *expected free energy* by following policies associated with minimal ambiguity and risk. Expected free energy then extends Active Inference to prospective and counterfactual forms of inference. This completes our journey along the low road to Active Inference. In chapter 3, we will travel the high road, which reaches the same conclusion on the basis of first principles and self-organization.

3 The High Road to Active Inference

Survival machines that can simulate the future are one jump ahead of survival machines who can only learn on the basis of overt trial and error. The trouble with overt trial is that it takes time and energy. The trouble with overt error is that it is often fatal. Simulation is both safer and faster.

—Richard Dawkins

3.1 Introduction

In chapter 2, we motivated the introduction of free energy as a means of performing approximate Bayesian inference (i.e., the low road to Active Inference). Here, we introduce free energy from another perspective, that of the high road, which inverts that reasoning: it starts from first principles in statistical physics and the central imperative that organisms must maintain their existence—that is, avoid surprising states—and then introduces the minimization of free energy as a computationally tractable solution to this problem. The chapter discloses the formal equivalence between the minimization of variational free energy and the maximization of model evidence (or self-evidencing) in approximate Bayesian inference, revealing a connection between free energy and Bayesian perspectives on adaptive systems. Finally, it discusses how Active Inference provides a novel first principle perspective to understand (optimal) behavior.

Active Inference is a theory of how living organisms maintain their existence by minimizing surprise—or a tractable proxy to surprise, *variational free energy*—via perception and action. By starting from first principles, it advances a novel belief-based scheme to understand behavior and cognition, which has numerous empirical implications.

The high road to Active Inference starts from the premise that, to survive, any living organism has to maintain itself in a suitable set of preferred states, while avoiding other, dis-preferred states of the environment. These preferred states are first and foremost defined by niche-specific evolutionary adaptations. However, as we will see later, in advanced organisms these can also extend to learned cognitive goals. For example, to survive, a fish has to stay in a comfort zone that corresponds to a small subset of all the possible states of the universe: it has to stay in water. Similarly, a human has to ensure that their internal states (e.g., physiological variables like body temperature and heart rate) always remain within acceptable ranges—otherwise they will die (or more precisely will become something else, such as a corpse). This acceptable range or comfort zone stipulatively defines the characteristic states something has to be in to be that thing.

Living organisms resolve this fundamental biological problem by exerting *active control* over their states (e.g., of body temperature) at many levels, which range from automatic regulatory mechanisms such as sweating (physiology) to cognitive mechanisms such as buying and consuming a drink (psychology) to cultural practices such as distributing air conditioning systems (social sciences).

From a more formal perspective, Active Inference casts the biological problem of—or explanation for—survival as surprise minimization. This formulation rests on a technical definition of *surprising states* from information theory—essentially, surprising states index those outside the comfort zone of living organisms. It then proposes free energy minimization as a practical and biologically grounded way for organisms or adaptive systems to minimize the surprise of sensory encounters.

3.2 Markov Blankets

An important precondition for any adaptive system is that it must enjoy some separation and autonomy from the environment—without which it would simply dissipate, dissolve, and thereby succumb to environmental dynamics. In the absence of this separation, there would be no surprise to minimize; there must be something to be surprised and something to be surprised about. In other words, there are at least two things—system and environment—and these can be disambiguated from one another. A formal way to express a separation between a system and the rest of the environment is the statistical construct of a *Markov blanket* (Pearl 1988); see box 3.1.

Box 3.1
Markov blankets

A Markov blanket is an important recurring concept in this book (Friston 2019a, Kirchhoff et al. 2018, Palacios et al. 2020). Technically, a blanket (b) is defined as follows:

$$\mu \perp x \,|\, b \Leftrightarrow p(\mu, x \,|\, b) = p(\mu \,|\, b)\, p(x \,|\, b)$$

This says (in two different but equivalent ways) that a variable μ is conditionally independent of a variable x if b is known. In other words, if we know b, knowing x would give us no additional information about μ. A common example of this is a Markov chain, where the past causes the present causes the future. In this scenario, the past may only influence the future via the present. This means no additional information about the future is gained by finding out about the past (assuming we know the present).

To identify a Markov blanket in a system wherein we know the conditional dependencies, we can follow a simple rule. The blanket for a given variable comprises its *parents* (the variables it depends on), its *children* (the variables that depend on it) and, in some settings, the other *parents of its children*.

In brief, a Markov blanket is the set of variables that mediate all (statistical) interactions between a system and its environment. Figure 3.1 illustrates an interpretation of a Markov blanket in a dynamic setting. Here the conditional independences have been supplemented with dynamical constraints, so that the flows do not depend upon states on the opposite side of the blanket.

The Markov blanket in figure 3.1 distinguishes states *internal* to the adaptive system (i.e., brain activity) from *external* states of the environment. Furthermore, it identifies two additional states, labeled *sensory states* and *active states*, which form the blanket that (statistically) separates internal and external states. Statistical separation means that if we knew about the active and sensory states, the external states would offer no additional information about internal states (and vice versa). In a dynamical setting, this is often interpreted as saying internal states cannot directly change external states but can do so vicariously by changing *active states*. Similarly, external states cannot directly change internal states but can do so indirectly by changing *sensory states*.

This is a restatement of the classical action-perception cycle, wherein an adaptive system and its environment can interact (only) through actions and observations, respectively. This reformulation has two main benefits.

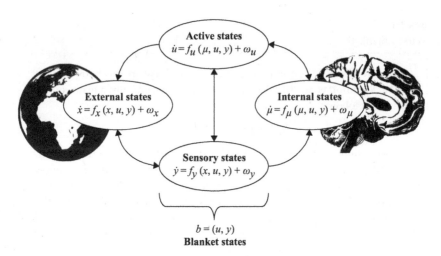

Figure 3.1
A dynamic Markov blanket, which separates an adaptive system (here, the brain) from the environment. The dynamics of each set of states are determined by a deterministic flow specified as a function (f) giving the average rate of change and additional stochastic (random) fluctuations (ω). The arrows indicate the direction of influence of each variable over the rates of change of other variables (technically, the nonzero elements of the associated Jacobians). This is just one example; one can use a Markov blanket to separate an entire organism from the environment or nest multiple Markov blankets within one another. For example, brains, organisms, dyads, and communities can be conceived in terms of different Markov blankets that are nested within one another (see Friston 2019a; Parr, Da Costa, and Friston 2020 for a formal treatment). Confusingly, different fields use different notations for the variables; sometimes, sensory states are denoted s, external states η, and active states a. Here we have chosen variables for consistency with the other chapters in this book.

First, it formalizes the fact that an adaptive system's internal states are autonomous from environmental dynamics and can therefore resist their influences. Second, it scaffolds the way in which adaptive systems minimize their surprise: it highlights the internal, sensory, and active states they have access to. Specifically, surprise is defined in relation to sensory states, while internal and active state dynamics are the means by which the surprise of sensory states may be minimized.

The key point to notice here is that the internal states of an adaptive system bear a formal relation to external states. This is due to a kind of symmetry across the Markov blanket as both influence and are influenced by blanket states. A consequence of this is that we can construct conditional probability distributions for the internal and external states, given the blanket states. Because these are conditioned on the same blanket states, we can associate pairs of expected internal and external states with one another. In other words, on average, the internal and external states acquire a kind of (generalized) synchrony—just as we might anticipate on attaching a pendulum to each end of a wooden beam. Over time, as they synchronize, each pendulum becomes predictive of the other through the vicarious influence of the beam (Huygens 1673). Figure 3.2 offers a graphical intuition for this relationship. This means that if we can write down independent

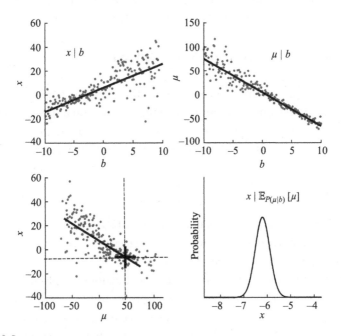

Figure 3.2

Association between average internal states of a Markov blanket and distributions of external states. *Top:* Assuming a linear Gaussian form for the conditional probabilities, these plots show samples from the conditional distribution over external and internal states, respectively, given blanket states. The thick black lines indicate the average of these variables given the associated blanket state. *Bottom left:* The same data are plotted to illustrate the synchronization of internal and external states afforded by sharing a Markov blanket—here, an inverse synchronization. The dashed lines and black cross illustrate that if we knew the average internal state (vertical line), we could identify the average external state (horizontal line) and the spread around this point. *Bottom right:* We can associate the average internal state with a distribution over the external state.

distributions over external and internal states given their Markov blanket, the two states become informative about one another via this blanket.

This synchrony gives internal states the appearance of representing (or *modeling*) external states—which links back to the idea of surprise minimization introduced in chapter 2. This is because surprise depends on an internal model of how sensory data are generated. To recap, minimizing the surprise (negative log probability) of sensory observations becomes identical to maximizing the evidence (marginal likelihood) for the model, which is just the probability of sensory observations under that model. This notion of surprise minimization can be understood from two equivalent— Bayesian and free energy—perspectives, which we discuss next.

3.3 Surprise Minimization and Self-Evidencing

Under a Bayesian perspective, an agent with a Markov blanket appears to *model* the external environment in the sense that internal states correspond (on average) to a probabilistic representation—an approximate *posterior belief*—of external states of the system (figure 3.2). The dynamics of internal states correspond to a form of (approximate) Bayesian inference of external states, as their motion changes the associated probability distribution, which is afforded by an implicit generative model of how sensations (or *sensory states* in the Markov blanket jargon) are generated. If we reinstate the notion of an agent as constituted by internal and blanket states, we can talk about an agent's generative model.

Importantly, the agent's generative model cannot simply mimic external dynamics (otherwise the agent would simply follow external dissipative dynamics). Rather, the model must also specify the preferred conditions for the agent's existence, or the regions of states that the agent has to visit to maintain its existence, or satisfy the criteria for its existence in terms of occupying characteristic states. These preferred states (or observations) can be specified as the *priors* of the model—which implies that the model implicitly assumes that its preferred (prior) sensations are more likely to occur (i.e., are less surprising) if it satisfies the criteria for existence. This means it has an implicit *optimism bias*. This optimism bias is necessary for the agent to go beyond the mere duplication of external dynamics to prescribe *active states* that underwrite its preferred or characteristic states.

Under this formulation, one can cast optimal behavior (with respect to prior preferences) as the *maximization of model evidence by perception and action*. Indeed, model evidence summarizes how well the generative model fits or explains sensations. A good fit indicates that the model success-fully accounts for its sensations (this is the *descriptive* side of inference); at the same time, it realizes its preferred sensations, given that they are less surprising (this is the *prescriptive* side of the inference). Such good fit is a guarantee of surprise minimization, as maximizing model evidence $P(y)$ is mathematically equivalent to minimizing surprise: $\mathfrak{I}(y) = -\ln P(y)$.

A way to reformulate the above arguments more succinctly consists in say-ing that any adaptive system engages in "self-evidencing" (Hohwy 2016). *Self-evidencing* here means acting to garner sensory data consistent with (i.e., that affords evidence to) an internal model, hence maximizing model evidence.

3.3.1 Surprise Minimization as a Hamiltonian Principle of Least Action

In the preceding sections, we have asserted that surprise must be minimized but have not detailed why this is. Although the details of the underlying phys-ics of self-evidencing are outside the scope of this book (see Friston 2019b for details), we here provide a brief overview of the principles. These are under-written by the idea that biological creatures—with Markov blankets—persist over time, resisting the dispersive effects of environmental fluctuations. The persistence of a Markov blanket implies that the distribution of blanket states remains constant over time. Simply put, this means that any deviation of sensory (or active) states from regions that are highly probable under this distribution must be corrected by the average flow of states (which is just the deterministic part of the flow in figure 3.1). Expressing this as a physicist might, stochastic (random) systems at steady state engage in dynamics that (on average) descend an energy function (or Hamiltonian) that is interpre-table as a negative log evidence or surprise. This is like a ball rolling down a hill from high gravitational potential energy at the top of the hill to low energy in a basin. See figure 3.3.

For the system shown on the left of figure 3.3, every time a fluctua-tion causes a move to a less probable state, this is corrected by a move up the probability gradient, such that the system occupies probability-dense regions a greater proportion of the time. The key insight here is that this system maintains sensory states within a narrow range by minimizing sur-prise (on average)—in contrast to the system on the right, for which surprise grows indefinitely.

Surprise minimization permits living organisms to (temporarily) resist the second law of thermodynamics, which states that entropy—or the

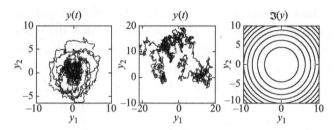

Figure 3.3

Left: Path taken by a 2-dimensional random dynamical system with a (nonequilibrium[1]) steady state. This can be interpreted as minimizing its surprise, which is shown in the contour plot on the right. *Right:* The center is the least surprising region; the circles moving away from the center represent progressively more surprising regions. *Middle:* In contrast, this plot shows the trajectory of a system starting in the same place (5, 5), with random fluctuations of the same amplitude, whose dynamics bear no relation to surprise. Not only does it enter more surprising regions of space; it also fails to achieve any sort of steady state, dissipating in an unconstrained fashion over time. The scope of Active Inference is restricted to systems like that on the left—which counter random fluctuations with their average flow and thereby retain their form over time.

dispersion of systemic states—always grows. This is because, on average, entropy is the long-term average of surprise and, on average, the maximization of a log probability of observations is equivalent to minimization of (Shannon) entropy:[2]

$$H[P(y)] = \mathbb{E}_{P(y)}[\Im(y)] = -\mathbb{E}_{P(y)}[\ln P(y)] \tag{3.1}$$

Ensuring that a small proportion of sensory states is occupied with high probability is equivalent to maintaining a particular entropy. This is a defining characteristic of self-organizing systems, as long recognized by cybernetic theories.

From a physiologist's perspective, surprise minimization formalizes the idea of homeostasis. As a sensor value leaves its optimal range, negative feedback mechanisms kick in that reverse these deviations. From a control perspective, we can interpret optimal behavior in relation to some desired steady state probability density. In other words, if we define a distribution of preferred outcomes, optimal behavior will involve evolution of the system toward—and maintenance of—that distribution.

As we saw in chapter 2, free energy is an upper bound on surprise, suggesting that optimal behavior can be obtained by minimizing free energy in the face of random fluctuations. Recall that the difference between free energy and surprise is the divergence between an exact posterior probability (i.e., the distribution of external states given blanket states) and an approximate posterior probability (i.e., the distribution over external states given average internal states). As such, the motion of internal states can be thought of as minimizing the divergence, which then enables active states, on average, to minimize the surprise accompanying sensory states. In other words, the optimal behavior resulting from free energy minimization is the one that is least surprising and follows a path of least Action[3] from the current state to the desired state—that is, the Hamiltonian principle of least Action applied to behavior.

Figure 3.3 shows a very simple example of a system equipped with a random attractor. This is analogous to a thermostat, which (in cybernetic parlance) has a single set-point and cannot learn or plan. Active Inference aims to use the same explanatory apparatus to cover much more complex and adaptive systems. Here, the difference between simplest and more complex systems can be reduced to the different shapes of their attractors—from fixed points to increasingly more complex and itinerant dynamics. From this perspective, one can understand living organisms as constantly seeking a compromise between excessive stability and excessive dispersion—and Active Inference aims to explain how such compromise is achieved.

3.4 Relations between Inference, Cognition, and Stochastic Dynamics

The physicist E. T. Jaynes famously argued that inference, information theory, and statistical physics are different perspectives on the same thing (Jaynes 1957). In the previous sections, we discussed how Bayesian and statistical physics perspectives offer two equivalent ways to understand surprise minimization and optimal behavior—effectively adding a form of cognition to Jaynes's triad. This equivalence between various schools of thought is appealing but can be confusing to those who are not familiar with the respective formalisms, where many different words are used to refer to the same quantities. To help demystify this, in this section we elaborate on the main equivalences between Bayesian and statistical physics perspectives and their cognitive interpretations; see table 3.1 for a summary and box 3.2.

Table 3.1

Statistical physics, Bayesian inference, and information theory—and their cognitive interpretations

Statistical physics	Bayesian inference and information theory	Cognitive interpretation
Minimize variational free energy	Maximize model evidence (or marginal likelihood); minimize surprisal (or self-information)	Perception and action
Minimize expected free energy; Hamiltonian principle of least Action	Infer the most likely (or less surprising) course of action	Planning as inference
Attain nonequilibrium steady-state	Perform approximate Bayesian inference	Self-evidencing
Gradient flows on energy functions; gradient descent on free energy	Gradient ascent on model evidence; gradient descent on surprisal	Neuronal dynamics

Box 3.2

Free energy in statistical physics and Active Inference

The notion of free energy is widely used in statistical physics to characterize (for example) thermodynamic systems. Although Active Inference uses exactly the same equations, it applies them to characterize the *belief state* of an agent (in relation to a generative model). Hence, when we talk of an Active Inference agent minimizing its (variational) free energy, we are referring to processes that change its belief state, not (for example) the particles of its body. To avoid misunderstandings, we use the term *variational free energy*, hence adopting a terminology that is more common in machine learning. Another more subtle point is that the concept of free energy is often used in the context of equilibrium statistical thermodynamics. Active Inference targets living organisms—or nonequilibrium steady state systems that are open—that feature continuous, reciprocal exchanges with the environment. This is an exciting novel field (Friston 2019a).

3.4.1   Variational Free Energy, Model Evidence, and Surprise

A first important equivalence is between the *maximization of model evidence* (or marginal likelihood) in Bayesian inference and the *minimization of variational free energy*—both of which minimize surprise. This equivalence becomes evident when one appeals to a specific approximate solution to intractable problems of inference—variational inference. Variational inference recasts the inference problem as an optimization problem by minimizing free energy. The minimum of the free energy is the point at which the approximation of the exact solution is at its best. Expressing this formally sheds light on the relations between the three quantities:

$$\underbrace{\Im(y\,|\,m)}_{\text{Surprise}} = \underbrace{-\ln P(y\,|\,m)}_{\text{Model evidence}} \leq \underbrace{D_{KL}\big[Q(x)\,\|\,P(x\,|\,y,m)\big] - \ln P(y\,|\,m)}_{\text{Variational free energy}} \qquad (3.2)$$

In equation 3.2, unlike in chapter 2, we have explicitly conditioned all quantities on a model, m, to emphasize that these depend on the model we have (or are) about how y are generated, and the quantities will vary if different models are used. The equivalence of these quantities raises the question as to why it is useful to distinguish between them. The main reason is that, unlike model evidence, variational free energy can be minimized efficiently.

Recall from chapter 2 that the variational free energy is only exactly equivalent to the negative model evidence or surprise when the KL-Divergence term becomes zero. This is not always possible, but this can be made close to zero. Hence, in the process of finding better and better values for $Q(x)$, variational free energy also approximates surprise more closely. We have said this a few times already because it is important to emphasize the central relationship between free energy and surprise that is the foundation of this book. Specifically, free energy is an *upper bound* on surprise. It can be the same as or greater than surprise—where what is *greater than* is quantified by the KL-Divergence.

An interesting aspect of this is that any system minimizing its surprise, including the very simple system in figure 3.2, is also minimizing a free energy, where the $Q(x)$ is always set to be equal to the exact posterior probability—that is, setting the KL-Divergence to be zero. One perspective on the difference between cognitive and noncognitive systems is that the latter always have a zero KL-Divergence, while cognitive systems must go through the (perceptual) process of minimizing this term before their actions are guaranteed to minimize surprise. Note that minimizing the divergence is the only thing that perception can do. This places a great deal of emphasis on the motion of internal states, such that the distribution they parameterize (figure 3.2) is as close to the exact posterior as possible. However, perception cannot minimize the second (evidence) component of variational free energy that corresponds to the actual surprise, because it cannot change the sensations that have been gathered. Only by acting in ways that change sensations can an agent minimize the second (evidence) component of variational free energy and resolve its surprise—or, equivalently, maximize its model evidence. This places emphasis on the motion of active states, given internal states, in self-evidencing.

An example helps in illustrating this point. Imagine that your generative model predicts a distribution of glucose levels in your blood given levels of hunger, with relatively high versus low glucose levels relating to satiation and hunger, respectively. In addition, imagine this model ascribes a higher prior probability to satiation and therefore to relatively high glucose levels— making low glucose levels surprising. Imagine you are initially uncertain about your hunger levels and sense low blood glucose. Perception leads to the inference that you are hungry and the experience of hunger—closing the KL-Divergence. However, perception cannot go further than that to reduce your surprise—and the discrepancy between the high level of glucose that you expect a priori and the low level of glucose that you sense—because it cannot act on your sensations (low glucose) or their causes (physiology). You can only minimize your surprise by acting to change (the hidden source of) the sensations you gather—for example, by eating a dessert.

In sum, perception can minimize variational free energy by reducing the discrepancy between approximate and true posterior but cannot go further in minimizing surprise. The next step of surprise minimization entails changing the sensations one gathers by acting, which is where inference goes beyond perception and becomes *active*.

3.4.2 Expected Free Energy and Inference of the Most Likely Trajectory

Another important equivalence is between the minimization of *expected free energy* and inferring the *most likely course of action*, or *policy*. This goes beyond specifying the least surprising part of state-space and deals with how surprising alternative routes to that part or location may be. These alternative paths are expressed in terms of policies, which are essentially trajectories across states. Importantly, in Active Inference the log probability of a policy is set proportional to the expected free energy if that policy was pursued. This implies that the most probable or least surprising path is (set to be) the one that minimizes expected free energy. This formulation is equivalent to the way *Action* is defined in physics, where it scores the probability of a path by an integral (or sum) of an energy. While a physical system may pursue a space of hypothetical trajectories, the path it actually follows is the one for which Action is minimized—that is, Hamilton's principle of least Action. This analogy between Active Inference and Hamilton's principle of least Action is unpacked in the next section.

3.5   Active Inference: A Novel Foundation to Understand Behavior and Cognition

In fields like optimal control, reinforcement learning, and economics, the optimization of behavior results from a *value function* of states, following Bellman's equation (Sutton and Barto 1998). Essentially, each state (or state-action pair) is assigned a value, which represents how good a state is for an agent to be in. The value of states (or state-action pairs) is usually learned by trial and error, by counting how many times—and after how much time—one obtains reward by starting from those states. Behavior consists in optimizing reward acquisition by reaching high-valued states, hence capitalizing on learning history.

In contrast, in Active Inference, behavior is the result of inference and its optimization is a function of beliefs. This formulation unites notions of (prior) belief and preference. As discussed above, using the notion of expected free energy amounts to endowing the agent with an implicit prior belief that it will realize its preferences. Hence, the agent's preference for a course of action becomes simply a belief about what it expects to do, and to encounter, in the future—or a belief about future trajectories of states that it will visit. This replaces the notion of value with the notion of (prior) belief. This is an apparently strange move, if one has a background in reinforcement learning (where value and belief are separated) or Bayesian statistics (where belief does not entail any value). However, it is a powerful move, for at least three reasons.

First, it automatically entails a self-consistent process model of purposive (or teleological) behavior, which is akin to cybernetic formulations. If we endow an Active Inference agent with some prior preference, then it will act to realize such preferences—because this is the only course of action consistent with its prior belief that it will act to fulfill its expectations. Note that the resulting (preferred) course of action, or policy, is directly measurable in experimental settings, whereas a value function or prior belief needs to be inferred and hence is a more indirect, if not tautological, measure.

Second, casting behavior as a functional of beliefs (probability distributions) automatically entails notions such as *degree of belief* and *uncertainty*. These notions undergird important facets of adaptive action but are not directly available in the Bellman formulation. By the same token, this formulation gives more flexibility in modeling sequential dynamics and itinerant behaviors, which are harder to model in terms of a value function of states (Friston, Daunizeau, and Kiebel 2009).

Third, in this formulation, optimal behavior comes to follow a Hamiltonian principle of least Action in statistical physics. Indeed, Active Inference goes one step further toward the idea that behavior is a function of beliefs: it also assumes that it becomes an energy function—and the most likely course of action of an Active Inference agent is the one that minimizes free energy. A profound consequence is that living organisms behave according to Hamilton's principle of least Action: they follow a path of least resistance until they reach a steady state (or a trajectory of states), as exemplified by the behavior of a random dynamical system (shown in figure 3.3). This is a fundamental assumption that distinguishes Active Inference from alternative theories of behavior and cognition based on the Bellman formulation.

It is worth briefly outlining what we mean by drawing analogies between Hamiltonian physics and Active Inference. This is intended on three levels. The first is that the advance offered by Active Inference to the behavioral and life sciences is comparable to the advance Lagrangian[4] and Hamiltonian formulations offered to Newton's accounts of mechanics. While Newtonian mechanics were originally formulated in terms of differential equations— including Newton's famous third law expressing the proportionality between acceleration and force—a complementary perspective on mechanics was offered by considering what is conserved by dynamical systems. Newtonian dynamics can then be derived from these conservation laws. These offer a perspective on which to base further theoretical advances, and they form the basis for parts of stochastic, relativistic, and quantum physics. Analogously, Active Inference reformulates the sorts of neuronal and behavioral dynamics that might previously have been built up from a series of differential equations by specifying the quantity—free energy—from which these dynamics may be derived. Just as different sorts of Hamiltonians lead to different types of physics, free energies based on different generative models lead to different neuronal and behavioral dynamics.

The second point of connection between Hamiltonian physics and Active Inference arises from a more direct association between a Hamiltonian and probability measures. The idea here is to associate the conserved Hamiltonian with the energy of the system. Remember that the quantities we have referred to as *energies* so far (here and in chapter 2) have all had the form of a negative log probability. This reflects an interpretation of energy as simply a measure of the improbability of any given configuration of a system. On this view, conservation of energy and of probability are equivalent laws. As dissipative systems—coupled to external states via a Markov blanket—move to states of low energy or high probability, we can directly associate the energy or Hamiltonian with surprise. As such, Active Inference *is* Hamiltonian physics applied to a certain kind of system (systems that feature a Markov blanket).

The third association between these formulations is the variational calculus that underwrites the association between energies and dynamics. This is most apparent when Hamiltonian physics is expressed as a principle of least Action, where *Action* refers to the integral of a Lagrangian over a path. Crucially, this Action is a functional of a path. Here, a path is a function of time whose output is the position and velocity of a particle on that path at that time. The path followed by a (deterministic) particle minimizes this Action. Similarly, Active Inference is predicated on the idea that beliefs (themselves functions of hidden states) must minimize a free energy functional. The key point of contact here is that in both cases, functions (paths or beliefs) must be optimized in relation to functionals (Action or free energy, respectively). This places both in the context of variational calculus, which is a branch of mathematics dedicated to finding extrema of functionals. In physics, this leads to the Euler-Lagrange equations. In Active Inference, we arrive at variational inference procedures.

3.6 Models, Policies, and Trajectories

In section 3.2, we highlighted that the scope of Active Inference pertains to those systems that enjoy some separation from their environment and saw that this translates into the presence of a Markov blanket. In section 3.3, we highlighted that the persistence of this blanket requires dynamics that (on average) minimize the surprise of (sensory) states. As this may be interpreted as self-evidencing, we arrive at the conclusion that behavior is determined by a steady-state distribution that can be interpreted as a generative model of how (sensory) data are generated.

This tells us something very important. Each generative model should be associated with different sorts of behavior. As such, different sorts of behavior may be accounted for by specifying different generative models—and implicitly what that system would find surprising. Furthermore, different kinds of generative model may correspond to adaptive or cognitive creatures having various levels of complexity (Corcoran et al. 2020). Very simple generative models of the sort driving the dynamics in figure 3.3 offer a minimal sort of cognition, as they cannot entertain the possibility of alternative (or counterfactual) trajectories. Further, these models are shallow, in the sense that they afford inference at just one timescale. In contrast, *hierarchical* generative models afford inference at multiple timescales. In hierarchical or deep models, the dynamics at higher hierarchical levels generally encode things that change more slowly (e.g., the sentence I am reading) and that contextualize things that change faster (e.g., the word I am reading), which are represented at lower hierarchical levels (Kiebel et al. 2008; Friston, Parr, and de Vries 2017).

What do we need to include in a model to derive more complex behaviors of the sort we would associate with agency and sentient systems? One answer to this is the capacity to model alternative futures, or different ways in which events might play out—and to select among them. In turn, considering possible futures requires a generative model that has some temporal depth and explicitly represents the consequences of actions. Working this into the model will ensure behavior that conforms to the most likely of these futures. The (counterfactual) capacity to entertain these alternatives may be what separates the steady state associated with sentient systems from simpler creatures. When alternative futures pertain to things over which we have control, we refer to these as policies or plans. As we saw in chapter 2, one way of disambiguating between these plans is to incorporate a prior belief into a model that says that those policies with the lowest expected free energy are the most plausible. This offers a way of characterizing a certain kind of system with a Markov blanket at steady state—which seems to correspond well to systems like us.

3.7   Reconciliation of Enactive, Cybernetic, and Predictive Theories under Active Inference

By emphasizing free energy minimization, Active Inference unites and extends three apparently disconnected theoretical perspectives.

First, Active Inference is in keeping with enactive theories of life and cognition, which emphasize the self-organization of behavior and autopoietic interactions with the environment, which ensure that living organisms remain within acceptable bounds (Maturana and Varela 1980). Active Inference provides a formal framework explaining how living organisms manage to resist the dispersion of their states by self-organizing a statistical structure—the Markov blanket—that affords reciprocal exchanges between organism and environment while also separating (and in a sense protecting the integrity of) the organisms' states from external, environmental dynamics.

Second, Active Inference is in keeping with cybernetic theories, which describe behavior as purposive and teleological. *Teleology* means that behavior is internally regulated by a mechanism that continuously tests whether a goal is achieved and, if not, steers corrective actions (Rosenblueth et al. 1943, Wiener 1948, Ashby 1952, G. Miller et al. 1960, Powers 1973). Similarly, Active Inference agents use both perception and action to minimize the discrepancy between preferred and sensed states. Active Inference provides a normative and viable description of the minimization process by specifying that what is actually minimized is a statistical quantity that the agent can measure—variational free energy—which under certain conditions corresponds to a *prediction error*, or the difference between what is expected and what is sensed. This implies a formulation of cybernetic control as a prospective process—which leads us to the next point.

Third, Active Inference is in keeping with theories that describe control as a prospective process that rests on a model of the environment—possibly physically implemented in the brain (Craik 1943). Active Inference assumes that agents use a (generative) model to construct predictions that guide perception and action and to evaluate their future (and counterfactual) action possibilities. This assumption is coherent with the good regulator theorem (Conant and Ashby 1970), which says that any controller should have—or be—a good model of the environment. Active Inference reconciles these model-based perspectives on brain and behavior under a rigorous characterization in terms of (approximate) Bayesian inference and (variational and expected) free energy minimization. Furthermore, Active Inference is largely coherent with ideomotor theory (Herbart 1825, James 1890, Hoffmann 1993, Hommel et al. 2001), which states that action starts with an imaginative process, and it is a predictive representation (of action consequences) that triggers actions—not a stimulus, like in stimulus-response theory (Skinner 1938). Active Inference casts this idea in an inferential framework, in which an action stems from a belief (about the future); this has a number of implications, such as the fact that in order to trigger action, one has to temporarily attenuate sensory evidence (which would otherwise falsify the belief that triggers action) (H. Brown et al. 2013).

The reconciliation of these frameworks is interesting, as they are often considered at odds. For example, self-organization and teleology are often seen as incompatible in biology. Furthermore, enactive theories tend to de-emphasize representation and control, which is instead a central construct of most theories of model-based inference. Active Inference formalizes autopoietic dynamics of adaptive agents from an unusual angle, which simultaneously considers self-organization and prediction. By connecting different perspectives, Active Inference can potentially help us understand how they illuminate one another.

3.8 Active Inference, from the Emergence of Life to Agency

Active Inference starts from first principles and unfolds them to explain behavior and cognition expressed by the simplest to the most complex forms of adaptive and living systems. In the continuum between simpler and more complex creatures, Active Inference draws a line between those that minimize variational free energy and those that also minimize expected free energy.

Any adaptive system that actively samples sensations to minimize *variational free energy* is (equivalently) an agent that actively gathers evidence for its generative model, aka a *self-evidencing agent* (Hohwy 2016). These systems are able to avoid dissipation, self-regulate, and survive by achieving set-points provided by basic homeostatic processes. These systems can generate complex and diverse forms of behavior and can also have very high fitness levels (as is already apparent in the case of viruses). Some may have hierarchical generative models that permit inferring events that change at different timescales, from faster (at lower hierarchical levels) to slower (at higher levels)—and hence can develop sophisticated strategies to deal with what they experience. However, these creatures are also fundamentally limited because their generative models lack temporal depth—or the capacity to plan and consider the future explicitly (although they can do so implicitly, for example, as a result of genetic evolution)—and hence they always live in the present.

A generative model endowed with temporal depth opens the door to the minimization of expected free energy—or in psychological terms, planning. In Active Inference, this entails much more than increased adaptivity: it entails at least a primitive form of agency. For an adaptive system, minimizing expected free energy is equivalent to having the (implicit) prior that one is a free energy minimizing agent—but acts to minimize free energy in the future. When this (prior) belief enters the generative model, the adaptive system becomes able to form beliefs about how it should behave in the future and which trajectories it will pursue. In other words, it becomes able to *select among alternative futures* as opposed to simply selecting *how to deal with the sensed present*, as in the simplest agents described above. This temporal depth therefore translates into a psychological depth. To ask about the ways living creatures populate the continuum between the simplest and most complex adaptive systems—and what forms of Active Inference they can express—is an empirical question.

3.9 Summary

The main topics of this chapter can be summarized as follows: Living organisms have to ensure that they only visit their characteristic or preferred states. If one defines these preferred states as expected states, then one can say that living organisms must minimize the surprise of their sensory observations (and maintain an optimal entropy; see box 3.3).

Doing this requires agents to exercise some autonomy from environmental dynamics and to be equipped with a Markov blanket that separates (i.e., expresses a conditional independence between) their internal states and the external states of the environment. Agents within the Markov blanket can engage in reciprocal (action-perception) exchanges with the environment. These exchanges are formally described by the theory of Active Inference, where both perception and action minimize surprise. They can do so by being equipped with a probabilistic generative model of how their

Box 3.3
Entropy minimization and open-ended behavior

Active Inference is based on the premise that living organisms strive to maintain a relative order (or negative entropy), controllability and predictability, despite being immersed in an environment whose natural forces generate continuous fluctuations—and a never-ending threat of entropic erosion. The most basic manifestation of this active pursuance of order is physiological homeostasis, with critical physiological parameters that need to be kept within viable regions. However, minimizing entropy should not be equated with a rigid repertoire of responses (e.g., autonomic homeostatic responses) but rather the opposite, especially in advanced organisms. We can develop open-ended repertoires of novel behaviors to pursue our original homeostatic imperatives—for example, to produce and buy good wine to satisfy thirst and other needs. This is sometimes referred to as "allostasis" (Sterling 2012).

More broadly, we actively pursue some order and controllability per se, without necessary reference to a specific homeostatic imperative—perhaps because preserving order facilitates many such imperatives. We actively carve our ecological niches to render them more predictable and less surprising. This is evident in the ways we construct our physical spaces (e.g., refuges and cities that give shelter from uncontrolled natural forces) and cultural spaces (e.g., societies with laws and deontic norms that give shelter from anarchic social forces). In all these examples, we usually need to accept some short-term increase of entropy or surprise (e.g., when we build something new or shift social stances) to ensure their long-term decrease. This helps us understand how the basic requirement for surprise minimization is not at odds with but rather promotes the epistemic imperatives and novelty-seeking, curious, and exploratory behavior that we recognize as central to many species.

A first way epistemic imperatives become apparent is during the minimization of *variational free energy*. One of the ways to decompose free energy is to express it as a Gibbs energy expected under the approximate posterior minus the entropy of the approximate posterior. In other words, the agent is striving to *increase* entropy. While this seems paradoxical, the paradox disappears if one considers that this is the entropy of the agent's (approximate posterior) belief. This can be understood as the imperative to explain things as accurately as possible but also "keep options open" and avoid committing to any specific explanation unless this is necessary—that is, the *maximum entropy* principle (Jaynes 1957).

A second way epistemic dynamics become apparent is during the minimization of *expected free energy*, wherein—interestingly—there are two entropies with opposite signs. These include the posterior predictive entropy (how

Box 3.3 (continued)

uncertain I am about what outcomes I would encounter given a choice) that must be maximized—as for beliefs about states in the variational free energy—and the conditional entropy of outcomes given states (the ambiguity entailed by a policy) that must be minimized. While during the minimization of variational free energy the imperative is to *maximize* entropy of (present) beliefs, during the minimization of expected free energy the imperative is to select actions that *minimize* the ambiguity of (future) beliefs. This gives rise to epistemic, curious, novelty-seeking, and information-foraging behaviors, which support uncertainty resolution or improvement of the generative model—which in turn minimizes surprise in the long run (Seth 2013; Friston, Rigoli et al. 2015; Seth and Friston 2016; Schwartenbeck, Passecker et al. 2019).

sensory observations are generated. This model defines surprise—or better, a tractable proxy, variational free energy, which can be measured and minimized efficiently.

An Active Inference agent appears to perform (approximate) Bayesian inference under a generative model and to maximize evidence for its model—that is, it is a self-evidencing agent. The prospective bit of the inference is realized by selecting courses of actions or policies that are expected to minimize free energy in the future. This formalism leads to a novel view of (optimal) behavior in terms of the Hamiltonian principle of least Action—a (first) principle that connects Active Inference to the domains of statistical physics, thermodynamics, and nonequilibrium steady states.

4 The Generative Models of Active Inference

Everything should be made as simple as possible, but not simpler.
—Albert Einstein

4.1 Introduction

This chapter complements the preceding chapters' conceptual treatment of Active Inference with a more formal treatment. Specifically, it sets out the relationship between free energy and Bayesian inference, the form of the generative models typically used in Active Inference, and the dynamics obtained from minimizing free energy for these models. A key focus is on how time is represented in a generative model. We will see the distinction between generative models formulated in continuous time and those that treat time as a sequence of events. Finally, we set out the idea of inferential message passing, which underwrites prominent theories in neurobiology—including predictive coding.

4.2 From Bayesian Inference to Free Energy

In the preceding two chapters, we outlined some of the important connections between Active Inference and other established paradigms in the neurosciences. In chapter 2, we focused on the notion of *the Bayesian brain* (Knill and Pouget 2004, Doya 2007)—one of its closest relatives—which provides a useful way to think about some of the consequences of active inference from a more formal perspective. Specifically, it helps us frame the problems that an agent engaging in Active Inference must solve. Broadly, these are the problem of inferring states of the world (perception) and inferring a course of action (planning). While it is tempting to equate Bayes optimality with exact Bayesian inference, exact inference is generally computationally intractable or even infeasible. In cognitive psychology and artificial intelligence applications, it is common to consider bounded forms of inference and rationality. We highlighted some examples in chapter 3. Under a Bayesian framework, this translates into using approximate inference. These methods comprise sampling methods and variational methods—on which active inference is based. In this section, we recap the basic elements of Bayesian inference and its variational manifestations (Beal 2003, Wainwright and Jordan 2008). In doing so, we hope to provide some intuition for the role of *free energy* and to emphasize the importance of *generative models* in drawing inferences about the world.

This chapter is more technical than chapters 1–3, appealing to a little linear algebra, differentiation, and the Taylor series expansion. Those readers interested in the details or in need of a refresher may turn to the appendices for the requisite background. Those who do not want to delve into the theoretical underpinnings may skip this chapter. Throughout, we explain the key implications of each equation—so it should be possible to develop an understanding of the important conceptual points herein even without following the formal argument.

A good place to start is Bayes' theorem. Recall from chapter 2 that this theorem expresses an equality between the product of a prior and a likelihood and the product of a posterior and a marginal likelihood. This is reproduced in equation 4.1:

$$P(x)P(y|x) = P(x|y)P(y)$$
$$P(y) = \sum_x P(y,x) = \sum_x P(y|x)P(x) \tag{4.1}$$

The first line of equation 4.1 is Bayes' theorem. The second line shows that the marginal likelihood (or model evidence), $P(y)$, can be computed directly from the prior and likelihood.[1] This makes the point that the prior and likelihood—which together comprise the generative model—are sufficient for us to compute the model evidence and the posterior probability. Despite this, it is not always easy to do so. The summation (or integration, if dealing with continuous variables) in equation 4.1 can be computationally or analytically intractable. One way to resolve this—the starting point of variational inference—is to convert this potentially difficult integration problem into an optimization problem. To understand how this works, we need to appeal to *Jensen's inequality*, which says that "the log[2] of an average

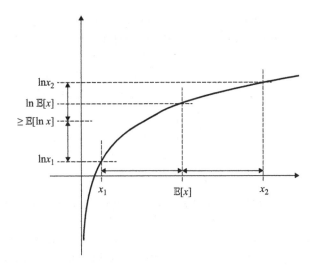

Figure 4.1
Logarithmic function providing intuition for Jensen's inequality. If we had only two data-points (x_1 and x_2), either we could take their average ($\mathbb{E}[x]$) and then find its log, or we could take the log of each data-point and then take the average of these ($\mathbb{E}[\ln x]$). The latter ($\mathbb{E}[\ln x]$) will always be below the former ($\ln \mathbb{E}[x]$), due to the concavity of the logarithmic function, unless the data-points are the same (where the log of the average and the average of the log are equal). This inequality holds for any number of data-points.

is always greater than or equal to the average of a log." Figure 4.1 provides a graphical intuition for why this is the case.

To take advantage of this property, we can rewrite equation 4.1 by multiplying the term inside the sum on the second line by an arbitrary function (Q) divided by itself (this is equivalent to multiplying by one, so the equality still holds) and taking the log of each side. Mathematically, this changes nothing. However, we can now interpret the expression as an expectation ($\mathbb{E}$)[3] of a ratio between two probabilities and so exploit Jensen's inequality:

$$\ln P(y) = \ln \sum_x P(y,x) \frac{Q(x)}{Q(x)}$$

$$= \ln \mathbb{E}_{Q(x)}\left[\frac{P(y,x)}{Q(x)}\right] \geq \mathbb{E}_{Q(x)}\left[\ln \frac{P(y,x)}{Q(x)}\right] \triangleq -F[Q,y] \tag{4.2}$$

The second line of this equation uses the fact that we have a log expectation and that, by Jensen's inequality, this must always be greater than or equal to the expectation of the log. This move is sometimes referred to as *importance sampling.* The right-hand side of this inequality is known as the (negative) variational free energy:[4] the smaller the free energy, the closer it is to the negative log model evidence. With this in mind, we can rewrite Bayes' theorem (equation 4.1) in logarithmic form, take its average under the posterior distribution, and disclose the relationship between this and the quantities of equation 4.2:

$$\ln P(x,y) = \ln P(y) \quad + \ln P(x|y) \Rightarrow$$
$$\mathbb{E}_{P(x|y)}[\ln P(x,y)] = \ln P(y) \quad + \mathbb{E}_{P(x|y)}[\ln P(x|y)] \tag{4.3}$$
$$\mathbb{E}_{Q(x)}[\ln P(x,y)] = -F[Q,y] + \mathbb{E}_{Q(x)}[\ln Q(x)]$$

The second line follows from the fact that the log probability of y is not a function of x, so taking an expectation under the posterior distribution does not change this quantity. Equation 4.3 provides some intuition for the roles of the free energy and the Q distribution—the two quantities that were difficult to compute without the variational approximation. The former plays the role of the negative log model evidence, while the latter acts as if it were the posterior probability. More formally, we can rearrange the free energy as we did in chapter 2 to quantify the relationship between free energy and model evidence:

$$F[Q,y] = \underbrace{D_{KL}[Q(x) \| P(x|y)]}_{\text{Divergence}} - \underbrace{\ln P(y)}_{\text{Log model evidence}}$$
$$D_{KL}[Q(x) \| P(x|y)] = \mathbb{E}_{Q(x)}[\ln Q(x) - \ln P(x|y)] \tag{4.4}$$

The first line of equation 4.4 shows the free energy expressed in terms of a KL-Divergence and a negative log evidence. The KL-Divergence is defined in the second line as the expected difference between two log probabilities. This is often used as a measure of how different two probability distributions are from one another.

Sometimes, the use of free energy is motivated directly in terms of this divergence. The argument goes that if our aim is to perform approximate Bayesian inference, we need to find an approximate posterior that best matches the exact posterior. As such, we can select a measure of the divergence between the two—of which the KL-Divergence in equation 4.4 is one example—and minimize this. As we do not know the exact posterior, we cannot use this divergence directly. One solution is to add the log evidence term, which may be combined with the log posterior to form the joint probability (which we do know because this is the generative model). The result is the free energy.

An interesting consequence of this perspective is that there is some ambiguity over which divergence measure to use. If we want to make the approximate and exact posterior as close as possible, we could use the other KL-Divergence, where Q and P are swapped, or choose from a large family of divergences, each of which emphasizes different aspects of the difference between distributions. However, the ideas set out in chapter 3 highlight the importance of self-evidencing for systems engaging in Active Inference. Therefore, we are primarily looking for a tractable evidence maximization scheme and only secondarily looking to minimize the divergence. From this perspective, there is no ambiguity as to which divergence measure to use. This emerges from the use of Jensen's inequality.

4.3 Generative Models

To calculate the free energy, we need three things: data, a family of variational distributions, and a generative model (comprising a prior and a likelihood). In this section, we outline two very general sorts of generative model used for Active Inference and the form the free energy takes in relation to each. The first deals with inferences about categorical variables (e.g., object identity) and is formulated as a sequence of events. The second deals with inferences about continuous variables (e.g., luminance contrast) and is formulated in continuous time using stochastic differential equations. Before specifying the details of these models, we review a graphical formalism that expresses the dependencies implied by a generative model.

Figure 4.2 shows several examples of generative models expressed as factor graphs, chosen to provide some intuition for the sorts of things that may be articulated in this way. These represent the factors (e.g., prior and likelihood) of a generative model as squares and the variables in that model (hidden states or data) in circles. Arrows indicate the direction of causality between these variables. The upper-left graph shows the simplest form these models can take, with a hidden state (x) causing data (y). The prior in this model is shown as factor 1, and the likelihood is factor 2. The other graphs extend this idea by introducing additional variables. In the upper right, z plays the role of a second hidden state, so that y depends on the states of both x and z.

As an example, consider a clinical diagnostic test. In this setting, the simple graph in the upper left can be interpreted as the presence or absence of a

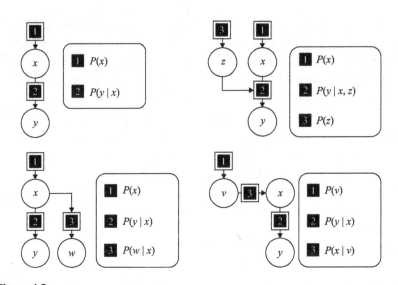

Figure 4.2
Dependencies between variables in a (graphical) probabilistic model. The circles represent random variables (i.e., the things about which we hold beliefs); the squares represent the probability distributions that describe the relationships between these variables. An arrow from one circle to another via a square indicates that the variable in the second circle depends on that in the first circle and that this dependency is captured in the probability distribution represented by the square.

disease (x) and the result of the test (y). The prior is then the prevalence of the disease, while the likelihood specifies the properties of the test. These include its specificity (the probability of a negative result in the absence of the disease) and sensitivity (the probability of a positive result in the presence of the disease). We can then think of the model in terms of the mechanism by which a test result is obtained—going from the top to the bottom of the factor graph. First, we sample a person from a population with known prevalence of a disease. If they have the disease, they will generate a true positive test result with probability given by the test sensitivity, and a false negative otherwise. If they do not have the disease, they will generate a true negative with probability given by the specificity, and a false positive otherwise.

Pursuing the same example, we can interpret the other factor graphs. In the upper-right panel, x and z could be the presence or absence of two different diseases, either of which could give a positive test result. In the lower left, w plays the role of data. Both y and w are generated by x and could represent (for example) two different diagnostic tests that are informative about the same disease process. Finally, the lower-right graph treats both x and v as hidden states but introduces a hierarchical structure in which v causes x causes y. Here we could think of v as providing a context or a pre-disposing factor (e.g., genetic polymorphism) for the presence or absence of disease x, which may be tested for by measuring y. In principle, we can add an arbitrary number of variables to this hierarchy.

Generative models of this sort are often used for static perceptual tasks, such as object recognition or cue integration. The generative models used for active inference differ in an important way: they evolve over time as new observations are sampled, and the observations that are added depend (via action) on beliefs about variables in the model. This has two key implications. First, the conditional dependencies include the dependencies of hidden variables at a given time on those at previous times. Second, these models sometimes include hypotheses about "how I am acting" as hidden variables.

Figure 4.3 illustrates the two basic forms of dynamic generative model used in active inference (Friston, Parr, and de Vries 2017) in factor graph form (Loeliger 2004, Loeliger et al. 2007). The upper graph shows a Partially Observable Markov Decision Process (POMDP), which expresses a model in which a sequence of states (s) evolves over time. At each time step, the current state is conditionally dependent on the state at the previous time and on the policy (π) currently being pursued. Policies here may be thought of as indexing alternative trajectories, or sequences of actions, that could be followed. Each time-point is associated with an observation (o) that depends only on the state at that time. This sort of model is very useful in dealing with sequential planning tasks—for example, navigating a maze (Kaplan and Friston 2018)—or decision-making processes that involve selecting between alternatives (e.g., categorization of a scene [Mirza et al. 2016]).

The lower graph in figure 4.3 shows a very similar graphical model but expressed in continuous time. In place of representing a trajectory as a series of states, this model represents the current position, velocity, and acceleration (and successive temporal derivatives) of a state (x). These values (referred to as *generalized coordinates of motion*) can be used to reconstruct a trajectory using a Taylor series expansion (see appendix A for an introduction to Taylor series approximations in this context). The relationship between a state and its temporal derivative here depends on (slowly varying) causes (v) that play a similar role to the policies above. As before,

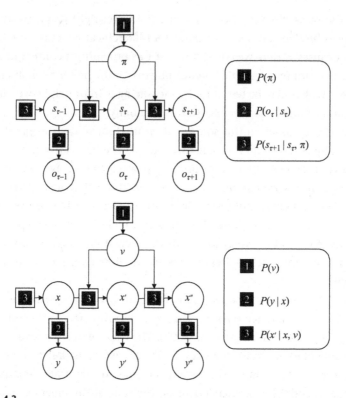

Figure 4.3
Two dynamic generative models (using the same graphical notation as in figure 4.2)
that we will appeal to throughout the remainder of this book. *Top:* Partially Observable Markov Decision Process (POMDP), defined in terms of a sequence of states
evolving through time (indexed by the subscript). *Bottom:* Continuous-time model,
of the sort implied by stochastic differential equations (with the prime notation indicating temporal derivatives).

states generate observations (y). The difference in notation (s, π, o vs. x, v, y) is
used to emphasize the difference between categorical variables that evolve
in discrete time and continuous variables that evolve in continuous time.
Similarly, from here on, we will use lowercase p and q for probability densities over continuous variables and uppercase P and Q for distributions
over categorical variables. Sections 4.4 and 4.5 will unpack these models
in more detail and will show how minimization of free energy in each
case leads to a set of equations that describes the dynamics of inferential
processes.

4.4   Active Inference in Discrete Time

In this section, we focus on the discrete-time model outlined above. This is important for understanding a range of cognitive processes that deal with categorical inferences and selection between alternative hypotheses. This formalism additionally facilitates an examination of the classic exploitation-exploration problem and illustrates how active inference resolves this.

4.4.1   Partially Observable Markov Decision Processes

As shown in figure 4.3, a POMDP expresses the evolution over time of a sequence of hidden states that depend on a policy. To specify this process formally, we need to account for the form of each of the square factor nodes in the figure. First, we describe each of these factors. We then combine them to express the joint distribution that constitutes the generative model.

As with the simple example of Bayes' rule given in chapter 2, we can separate the factors into those representing a likelihood and those combining to make a prior. The likelihood is similar to that used before and expresses the probability of an outcome (observable) given a state (hidden). If both the outcomes and states are categorical variables, the likelihood is a categorical distribution, parameterized by a matrix, $\mathbf{A}$:

$$P(o_\tau | s_\tau) = Cat(\mathbf{A})$$
$$A_{ij} = P(o_\tau = i | s_\tau = j) \tag{4.5}$$

The second line here details what is meant by the Cat notation (i.e., specification of a categorical distribution). This accounts for the nodes labeled "2" in figure 4.3. The prior over the sequence (expressed using the $\sim$ symbol) of hidden states depends on two things: the prior over the initial state (specified by a vector, $\mathbf{D}$) and beliefs about how the state at one time transitions to that at the next (specified as a matrix, $\mathbf{B}$):

$$P(\tilde{s} | \pi) = P(s_1) \prod_{\tau=1} P(s_{\tau+1} | s_\tau, \pi)$$
$$P(s_1) = Cat(\mathbf{D}) \tag{4.6}$$
$$P(s_{\tau+1} | s_\tau, \pi) = Cat(\mathbf{B}_{\pi\tau})$$

Together, these account for the "3" nodes in figure 4.3. Note that the transitions are conditionally dependent on the policy chosen. Thus, we can interpret the priors of equation 4.6, combined with the likelihood of equation 4.5, as expressing a model (π) of a behavioral sequence. To allow us to select between these models (i.e., to form a plan), we need a prior belief about the most probable sequence. For a free energy minimizing creature, a self-consistent prior is that the most probable policies are those that will lead to the lowest expected free energy (G) in the future:

$$P(\pi) = Cat(\boldsymbol{\pi}_0)$$
$$\boldsymbol{\pi}_0 = \sigma(-\mathbf{G})$$
$$\mathbf{G}_\pi = G(\pi) = -\mathbb{E}_{\tilde{Q}}[D_{KL}[Q(\tilde{s}|\tilde{o},\pi)\,\|\,Q(\tilde{s}|\pi)]] - \mathbb{E}_{\tilde{Q}}[\ln P(\tilde{o}|C)] \qquad (4.7)$$
$$\tilde{Q}(o_\tau, s_\tau|\pi) \triangleq P(o_\tau|s_\tau)Q(s_\tau|\pi)$$

This equation, being of fundamental importance to Active Inference, is worth unpacking in more depth. The first two lines express the prior probability for each policy, as parameterized by π_0, as being related to the negative expected free energy associated with that policy. The softmax function (σ) enforces normalization (i.e., ensures that the probability over policies sums to one). The final two lines of equation 4.7 express the form of the expected free energy.

Note the similarity between this and the functional form of the free energy (equation 4.4)—with a log probability of outcomes and a KL-Divergence. The key difference here is that the expectation is taken with respect to the *posterior predictive* density as defined by the final equality. This distribution expresses a joint probability over future states and observations. Crucially, this means we can compute the expected free energy in the future—something we could not do with the variational free energy, which depends on (present and past) observations. In addition, note the distribution over outcomes depends on parameters (C) and the reversal of the sign of the KL-Divergence, which is a consequence of the expectation under the posterior predictive probability. This last point can cause some confusion, so it is worth spelling out explicitly why this is. In the context of the variational free energy, the KL-Divergence was the expected difference between the log probability of the approximate posterior and the log probability of the exact posterior (equation 4.4). The analogous term in the expected free energy is the expected difference between the approximate posterior and the exact posterior we would get on the basis of the entire trajectory of outcomes, using current posterior beliefs as if they were priors. Unpacking this, we get the following:

$$\mathbb{E}_{\tilde{Q}}\big[\ln Q(\tilde{s}|\pi) - \ln Q(\tilde{s}|\tilde{o},\pi)\big]$$
$$= \mathbb{E}_{Q(\tilde{o}|\pi)}\Big[\mathbb{E}_{Q(\tilde{s}|\tilde{o},\pi)}\big[\ln Q(\tilde{s}|\pi) - \ln Q(\tilde{s}|\tilde{o},\pi)\big]\Big]$$
$$= -\mathbb{E}_{Q(\tilde{o}|\pi)}\Big[\mathbb{E}_{Q(\tilde{s}|\tilde{o},\pi)}\big[\ln Q(\tilde{s}|\tilde{o},\pi) - \ln Q(\tilde{s}|\pi)\big]\Big] \qquad (4.8)$$
$$= -\mathbb{E}_{Q(\tilde{o}|\pi)}\big[D_{KL}[Q(\tilde{s}|\tilde{o},\pi)\,\|\,Q(\tilde{s}|\pi)]\big]$$

Here we see that the order in which we must take expectations is important. It prompts a reversal in sign relative to the analogous term in the variational free energy. This underwrites an important difference between the two quantities. The expected free energy is minimized by selecting those observations that cause a large change in beliefs, in contrast to the variational free energy that is minimized when observations comply with current beliefs. This is the difference between optimizing beliefs in relation to data that have already been gathered (variational free energy minimization) and selecting those data that will best optimize beliefs (expected free energy minimization).

This reiterates that Active Inference uses two constructs, variational free energy (F) and expected free energy (G), which are mathematically related but play distinct and complementary roles. Variational free energy is the primary quantity that is minimized over time. It is optimized in relation to a generative model, which can include policies (or action sequences). As with all other hidden states, the agent needs to assign a prior probability to policies—because policies are just another random variable in the generative model. Active Inference uses a prior that is (loosely speaking) equivalent to the belief that one will minimize free energy in the future: that is, the expected free energy. In other words, expected free energy furnishes a prior over policies and is therefore a prerequisite in minimizing variational free energy.

In chapter 2 we saw that, as with the variational free energy, the expected free energy can be rearranged in a number of ways to disclose various interpretations. Here, we focus on an interpretation as the difference between the *risk* and the *ambiguity* associated with a policy. This is equivalent to the expression in equation 4.7:

$$G(\pi) = \underbrace{-\mathbb{E}_{\tilde{Q}}[D_{KL}[Q(\tilde{s}|\tilde{o},\pi)\,||\,Q(\tilde{s}|\pi)]]}_{\text{Information gain}} \underbrace{-\,\mathbb{E}_{\tilde{Q}}[\ln P(\tilde{o}|C)]}_{\text{Pragmatic value}}$$
$$= \underbrace{\mathbb{E}_{\tilde{Q}}[H[P(\tilde{o}|\tilde{s})]]}_{\text{Expected ambiguity}} + \underbrace{D_{KL}[Q(\tilde{o}|\pi)\,||\,P(\tilde{o}|C)]}_{\text{Risk}} \tag{4.9}$$

Recall from chapter 2 that the first of these expresses the trade-off between seeking new information (i.e., exploration) and seeking preferred observations (i.e., exploitation). By minimizing expected free energy, the relative balance between these terms determines whether behavior is predominantly explorative or exploitative. Note that pragmatic value emerges as a prior belief about observations, where the C-parameters of this distribution may be chosen to reflect the sort of system we are interested in characterizing (in terms of its characteristic or preferred outcome states). Following the second line of equation 4.9, we can rewrite equation 4.7 in linear algebraic form as follows:

$$\boldsymbol{\pi}_0 = \sigma(-\mathbf{G})$$
$$\mathbf{G}_\pi = \mathbf{H} \cdot \mathbf{s}_{\pi\tau} + \mathbf{o}_{\pi\tau} \cdot \boldsymbol{\varsigma}_{\pi\tau}$$
$$\boldsymbol{\varsigma}_{\pi\tau} = \ln \mathbf{o}_{\pi\tau} - \ln \mathbf{C}_\tau$$
$$\mathbf{H} = -diag(\mathbf{A} \cdot \ln \mathbf{A})$$
$$P(o_\tau \mid C) = Cat(\mathbf{C}_\tau)$$
$$Q(o_\tau \mid \pi) = Cat(\mathbf{o}_{\pi\tau}), \quad \mathbf{o}_{\pi\tau} = \mathbf{A}\mathbf{s}_{\pi\tau}$$
$$Q(s_\tau \mid \pi) = Cat(\mathbf{s}_{\pi\tau})$$
$$Q(s_\tau) = Cat(\mathbf{s}_\tau), \quad \mathbf{s}_\tau = \sum_\pi \boldsymbol{\pi}_\pi \mathbf{s}_{\pi\tau}$$

(4.10)

The first line of equation 4.10 uses a softmax (normalized exponential) operator to construct a probability distribution (parameterized with sufficient statistics $\boldsymbol{\pi}_0$) that sums to one from the expected free energy vector. Lines two to four express the components of the expected free energy in linear algebraic notation. The fifth line shows that the prior belief about observations is a categorical distribution (whose sufficient statistics are given in the $\mathbf{C}$ vector). The sixth to eighth lines specify the relationship between the linear algebraic quantities and the associated probability distributions. Having completed the specification of the generative model, we can now express the free energy in terms of the variables above:

$$F = \boldsymbol{\pi} \cdot \mathbf{F}$$
$$\mathbf{F}_\pi = \sum_\tau \mathbf{F}_{\pi\tau}$$
$$\mathbf{F}_{\pi\tau} = \mathbf{s}_{\pi\tau} \cdot (\ln \mathbf{s}_{\pi\tau} - \ln \mathbf{A} \cdot o_\tau - \ln \mathbf{B}_{\pi\tau} \mathbf{s}_{\pi\tau-1})$$

(4.11)

The decomposition of this into a sum over time is due to the implicit meanfield approximation that assumes we can factorize the approximate posterior into a product of factors:

$$Q(\tilde{s} \mid \pi) = \prod_\tau Q(s_\tau \mid \pi)$$

(4.12)

In logarithmic form, this becomes a sum, just as in equation 4.11. This factorization is one of many possibilities in variational inference—and represents the simplest option. In practice, this is often nuanced slightly, as detailed in appendix B.

4.4.2 Active Inference in a POMDP

Hitherto, we have defined the four key ingredients for a discrete-time generative model. These are the likelihood (**A**), transition probabilities (**B**), prior beliefs about observations (**C**), and prior belief about the initial state (**D**). Once these probability distributions are specified, a generic message passing scheme can be employed to minimize free energy and solve the POMDP. To make inferences about hidden states under a given policy, we set the rate of change of an auxiliary variable (**v**), which stands in for the log posterior (**s**), to be equal to the negative free energy gradient. A softmax (normalized exponential) function is then used to compute **s** from **v**.

$$\mathbf{s}_{\pi\tau} = \sigma(\mathbf{v}_{\pi\tau})$$
$$\dot{\mathbf{v}}_{\pi\tau} = \boldsymbol{\varepsilon}_{\pi\tau} \triangleq -\nabla_s \mathbf{F}_{\pi\tau}$$
$$= \ln \mathbf{A} \cdot o_\tau + \ln \mathbf{B}_{\pi\tau} \mathbf{s}_{\pi\tau-1} + \ln \mathbf{B}_{\pi\tau+1} \cdot \mathbf{s}_{\pi\tau+1} - \ln \mathbf{s}_{\pi\tau} \tag{4.13}$$

Equation 4.13 can be regarded as an example of variational message passing (see box 4.1). To update beliefs about policies, we find the posterior that minimizes the free energy:

$$\nabla_\pi F = 0 \Leftrightarrow$$
$$\pi = \sigma(-\mathbf{G} - \mathbf{F}) \tag{4.14}$$

For the simplest form of POMDP, equations 4.13 and 4.14 can be used to solve an Active Inference problem for any set of probability matrices; these may be thought of as describing perception and planning, respectively. We will unpack this in greater detail in the second part of the book, where we will provide worked examples of Active Inference for perception and planning (and other cognitive functions).

Figure 4.4's graphical representations of equations 4.10, 4.13, and 4.14 hint at possible neuronal implementations of free energy minimization in the brain—if one interprets nodes as neuronal populations, edges as synapses, and messages as synaptic exchanges. In later chapters we will consider the extension of this to factorized state-spaces, deep temporal models, and the optimization of the parameters of the generative model itself (learning).

4.5 Active Inference in Continuous Time

In the previous section, we dealt with the form Active Inference takes under a particular choice of generative model. These POMDPs are a useful way to

Box 4.1

Message passing and inference

Markov blankets

We encountered the concept of a Markov blanket in chapter 3. However, it is worth briefly reviewing the idea here. It relates to a system of multiple interacting variables. A Markov blanket for a given variable comprises a subset of those that interact with it. If we know everything about this subset, knowledge of anything outside this subset does not increase our knowledge of the variable of interest. The relevance here is that we can draw inferences about a variable in a graphical model based on local information about its Markov blanket. The blanket of a variable x are those variables that cause x (*parents*, $\rho(x)$), the variables that are caused by x (*children*, $\kappa(x)$), and the parents of x's children. Using this notation, two of the most common Bayesian message passing schemes used for approximate inference are defined as follows:

Variational message passing

$$\ln Q(x) = \mathbb{E}_{Q(\rho(x))}[\ln P(x\,|\,\rho(x))] + \mathbb{E}_{\underbrace{Q(\kappa(x))Q(\rho(\kappa(x)))}_{Q(x)}}[\ln P(\kappa(x)\,|\,\rho(\kappa(x)))]$$

This involves messages from all constituents of the Markov blanket of x, including the parents (via the conditional probability of x given its parents) and the children. The latter depends on the conditional probability of the children of x given all of their parents—which include x. Note the expectation includes the children and parents of the children. As the parents of the children include x, we divide by $Q(x)$ to ensure the expectation includes the blanket only.

Belief propagation

$$\ln Q(x) = \ln \mu_\kappa(x) + \ln \mu_\rho(x)$$
$$\mu_\kappa(x) = \mathbb{E}_{\frac{\mu_\kappa(\kappa(x))\mu_\rho(\kappa(x))}{\mu_x(\kappa(x))}}[P(\kappa(x)\,|\,\rho(\kappa(x)))]$$
$$\mu_\rho(x) = \mathbb{E}_{\frac{\mu_\rho(\rho(x))\mu_\kappa(\rho(x))}{\mu_x(\rho(x))}}[P(x\,|\,\rho(x))]$$

This has broadly the same structure as variational message passing but uses a recursive definition of messages such that each message ($\mu_a(b)$ being the message to b from a) depends on other messages (the messages to a). There is a directional aspect to this, such that the message from a to b depends on all messages to a, except for that from b (hence the division in the expectations). NB: The slightly nonstandard use of the expectation operator here allows us to (1) cover both discrete and continuous variables and (2) highlight the formal similarity between variational message passing and belief propagation.

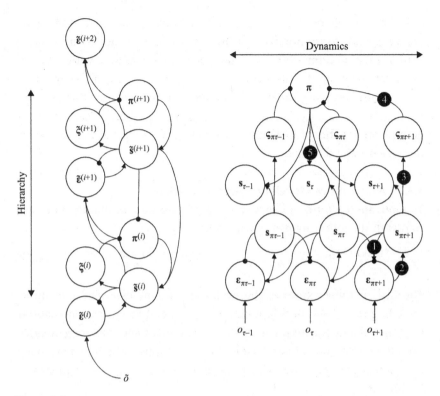

Figure 4.4

Bayesian message passing. *Right:* Dependencies between different variables in the belief-updating scheme outlined in the main text. Intuitively, current beliefs about states (under each policy) at each time are compared with those that would be predicted given beliefs about states at other times (1) and current outcomes to calculate prediction errors. These errors then drive updating in these beliefs (2); given beliefs about states under each policy, we can then calculate the gradients of the expected free energy (3). These are combined with the outcomes predicted under each policy (omitted from the figure) to compute beliefs about policies (4). Using a Bayesian model average, we can then compute posterior beliefs about states averaged over policies (5). This high-level summary of message passing omits some intermediate connections that could be included (e.g., connection (4) could be unpacked to explicitly include computation of the expected free energy). *Left:* This scheme could be expanded hierarchically (collapsing over time steps and policies for simplicity). The key idea is that a higher-level network might predict the states and policies at the lower level and use these to draw inferences about the context in which these occur. We will unpack this idea further in chapter 7.

articulate a range of inference problems, including those that underwrite planning and decision-making. However, when it comes to interacting with a real environment, models described in discrete time with categorical variables fall short. This is because sensory input and motor outputs are continuously evolving variables. To account for this, we now turn to a different sort of generative model. We apply exactly the same idea, a gradient descent on variational free energy, to these models to find the analogous message passing schemes.

4.5.1 A Generative Model for Predictive Coding

To motivate the form of generative model used for continuous states, we start with the following pair of equations:

$$\dot{x} = f(x,v) + \omega_x$$
$$y = g(x,v) + \omega_y \tag{4.15}$$

The first of these expresses the evolution of a hidden state over time, according to a deterministic function ($f(x,v)$) and stochastic fluctuations (ω). The second equation expresses the way in which data are generated from the hidden state. In each case, the fluctuations are assumed normally distributed, giving the following probability densities for the dynamics and likelihood:

$$p(\dot{x}|x,v) = \mathcal{N}(f(x,v),\Pi_x)$$
$$p(y|x,v) = \mathcal{N}(g(x,v),\Pi_y) \tag{4.16}$$

The precision (Π) terms are the inverse covariance of the fluctuations. These two equations form the generative model that underwrite Kalman-Bucy filters in engineering. However, schemes of this sort are limited by the assumption of uncorrelated fluctuations over time (i.e., Wiener assumptions). This is inappropriate for inference in biological systems, where fluctuations are themselves generated by dynamical systems and have a degree of smoothness. We can account for this by considering not only the rate of change of the hidden state and the current value of the data but also their velocities, accelerations, and subsequent temporal derivatives—that is, generalized coordinates of motion (Friston, Stephan et al. 2010; see box 4.2):

Box 4.2

Generalized coordinates of motion

To represent a trajectory in continuous time, generalized coordinates of motion provide a simple parameterization. This is based on a polynomial (Taylor series) expansion around the present time to give a function that lets us extrapolate to the recent past and near future. The plots in figure 4.5 show a trajectory in some space (x) over time (τ) as a solid line. From left to right, they show the trajectory represented in generalized coordinates of motion with one, two, and three coordinates (successive temporal derivatives of x). This is the dashed line. The expansion here is around the initial time point. With each successive generalized coordinate, we get a more accurate approximation of the trajectory into the proximal future. For most applications, around six generalized coordinates are sufficient.

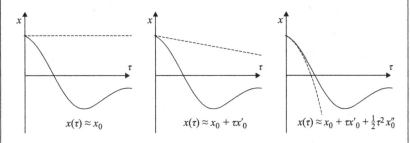

Figure 4.5

$$
\begin{aligned}
\dot{x} &= f(x,v) + \omega_x & y &= g(x,v) + \omega_y \\
\dot{x}' &= f'(x',v') + \omega_x' & y' &= g'(x',v') + \omega_y' \\
\dot{x}'' &= f''(x'',v'') + \omega_x'' & y'' &= g''(x'',v'') + \omega_y'' \\
&\vdots & &\vdots \\
\dot{x}^{[i]} &= f^{[i]}(x^{[i]},v^{[i]}) + \omega_x^{[i]} & y^{[i]} &= g^{[i]}(x^{[i]},v^{[i]}) + \omega_y^{[i]} \\
&\vdots & &\vdots
\end{aligned}
\tag{4.17}
$$

These generalized coordinates can be summarized more succinctly by representing a trajectory (again using the $\sim$ symbol) as a vector with elements corresponding to the successive derivatives above:

$$
\left.
\begin{aligned}
D\tilde{x} &= \tilde{f}(\tilde{x},\tilde{v}) + \tilde{\omega}_x \\
\tilde{y} &= \tilde{g}(\tilde{x},\tilde{v}) + \tilde{\omega}_y
\end{aligned}
\right\}
\Rightarrow
\begin{aligned}
p(\tilde{x}|\tilde{v}) &= \mathcal{N}(D \cdot \tilde{f}, \tilde{\Pi}_x) \\
p(\tilde{y}|\tilde{x},\tilde{v}) &= \mathcal{N}(\tilde{g}, \tilde{\Pi}_y)
\end{aligned}
\tag{4.18}
$$

In equation 4.18, D is a matrix with ones above the leading diagonal and zeros elsewhere. This effectively shifts all elements of the vector upward and may be thought of as a derivative operator. The generalized precision matrices may be constructed on the basis of the smoothness we assume for the fluctuations, as detailed in appendix B. Equipped with a prior over the hidden cause (v), whose relevance will become clearer below, this lets us write down the free energy for this generative model:

$$
\begin{aligned}
F[\mu, y] &= -\ln p(\tilde{y}, \tilde{\mu}_x, \tilde{\mu}_v) \\
&= \tfrac{1}{2} \tilde{\varepsilon} \cdot \tilde{\Pi} \tilde{\varepsilon} \\
&= \tfrac{1}{2} \left(\tilde{\varepsilon}_y \cdot \tilde{\Pi}_y \tilde{\varepsilon}_y + \tilde{\varepsilon}_x \cdot \tilde{\Pi}_x \tilde{\varepsilon}_x + \tilde{\varepsilon}_v \cdot \tilde{\Pi}_v \tilde{\varepsilon}_v \right)
\end{aligned}
$$

$$
\tilde{\varepsilon} = \begin{bmatrix} \tilde{\varepsilon}_y \\ \tilde{\varepsilon}_x \\ \tilde{\varepsilon}_v \end{bmatrix} = \begin{bmatrix} \tilde{y} - \tilde{g}(\tilde{\mu}_x, \tilde{\mu}_v) \\ D\tilde{\mu}_x - \tilde{f}(\tilde{\mu}_x, \tilde{\mu}_v) \\ \tilde{\mu}_v - \tilde{\eta} \end{bmatrix} \qquad (4.19)
$$

$$
\tilde{\Pi} = \begin{bmatrix} \tilde{\Pi}_y & & \\ & \tilde{\Pi}_x & \\ & & \tilde{\Pi}_v \end{bmatrix}
$$

In equation 4.19, the μ terms indicate the mode of the approximate posterior density for the x and v terms. The reason the free energy takes such a simple form in the first line is that we have employed a Laplace approximation, as detailed in box 4.3. In brief, this treats all probability densities as Gaussian, which—through a Taylor series expansion—is equivalent to assuming we are operating close to the mode of the distribution. The second line of the equation expresses the log probability in terms of squared precision weighted prediction errors. This omits all terms that are constant with respect to the posterior mode. The third line unpacks this in terms of the log likelihood, log probability of x given v, and log prior of v.

4.5.2 Active Inference as Predictive Coding with Motor Reflexes

Because the variance of the approximate posterior is an analytic function of the mode, under the Laplace approximation, we can optimize the free energy with respect to the mode. A simple way to think about this is that we need only find the *maximum a posteriori* (MAP) estimates[5] for each state. These are the means of the posterior distribution that may be equipped with its precision without need for further inference via the Laplace approximation (see box 4.3).

Box 4.3

The Laplace approximation

Laplace approximations rely on a principle similar to the generalized coordinates of motion described in box 4.2. The idea is that the free energy may be approximated by a quadratic expansion around the posterior mode (μ). In one dimension, this is as follows:

$$F[y, q] = \mathbb{E}_{q(x)}[\ln q(x) - \ln p(y, x)]$$

$$\approx \mathbb{E}_{q(x)}\left[\ln q(\mu) + (x - \mu)\underbrace{\partial_x \ln q(x)\big|_{x=\mu}}_{=0} + \tfrac{1}{2}(x - \mu)^2 \partial_x^2 \ln q(x)\big|_{x=\mu} \right.$$
$$\left. - \ln p(y, \mu) - (x - \mu)\partial_x \ln p(y, x)\big|_{x=\mu} - \tfrac{1}{2}(x - \mu)^2 \partial_x^2 \ln p(y, x)\big|_{x=\mu} \right]$$

The assumption that a quadratic expansion is sufficient is equivalent to saying that we can treat the probabilities as Gaussian (as the log of a Gaussian density is quadratic). Making this explicit, we can simplify the above to the following:

$$q(x) = \mathcal{N}(\mu, \Sigma^{-1})$$

$$F[y, \mu] = -\ln 2\pi \Sigma - \ln p(y, \mu) - \frac{1}{2} tr\left[\Sigma \partial_x^2 \ln p(y, x)\big|_{x=\mu} \right]$$

Under quadratic assumptions, the only term that depends on the mode is the second term. Omitting the other terms leads to the expression in equation 4.19. We can find the precision of the approximate posterior directly, once we know the mode, through the following expansion:

$$\ln q(x) \approx \ln p(x|y)$$
$$= \ln p(x, y) - \ln p(y)$$
$$\approx \ln p(\mu, y) + (x - \mu) \cdot \underbrace{\partial_x \ln p(x, y)\big|_{x=\mu}}_{=0}$$
$$+ \tfrac{1}{2}(x - \mu) \cdot \partial_x^2 \ln p(x, y)\big|_{x=\mu} (x - \mu) - \ln p(y)$$
$$\Rightarrow q(x) \propto e^{-\frac{1}{2}(x-\mu)\cdot\Sigma^{-1}(x-\mu)}, \quad \Sigma^{-1} = -\partial_x^2 \ln p(x, y)\big|_{x=\mu}$$

This tells us that the posterior precision is simply the second derivative of the joint probability evaluated at the posterior mode.

$$\dot{\mu} - D\tilde{\mu} = -\nabla_{\tilde{\mu}}F$$
$$= \nabla_{\tilde{\mu}} \ln p(\tilde{y}, \tilde{\mu})$$
$$= -\nabla_{\tilde{\mu}} \tilde{\varepsilon} \cdot \tilde{\Pi}\tilde{\varepsilon}$$

$$\begin{bmatrix} \dot{\mu}_x - D\tilde{\mu}_x \\ \dot{\mu}_v - D\tilde{\mu}_v \end{bmatrix} = \begin{bmatrix} \nabla_{\tilde{\mu}_x} \tilde{g} \cdot \tilde{\Pi}_y \tilde{\varepsilon}_y - D \cdot \tilde{\Pi}_x \tilde{\varepsilon}_x + \nabla_{\tilde{\mu}_x} \tilde{f} \cdot \tilde{\Pi}_x \tilde{\varepsilon}_x \\ \nabla_{\tilde{\mu}_v} \tilde{g} \cdot \tilde{\Pi}_y \tilde{\varepsilon}_y + \nabla_{\tilde{\mu}_v} \tilde{f} \cdot \tilde{\Pi}_x \tilde{\varepsilon}_x - \tilde{\Pi}_v \tilde{\varepsilon}_v \end{bmatrix}$$

(4.20)

In contrast to the gradient descents we saw for the discrete-time scheme, the left-hand side of equation 4.20 is the difference between the rate of change of μ and the derivative operator applied to this. This is because when the free energy is minimized, it does not make sense for the rate of change of the posterior mode to be zero if the posterior mode associated with rates of change is nonzero. In other words, "the motion of the mode should be the mode of the motion" at the free energy minimum. This ensures $\dot{\mu}^{[i]} = \mu^{[i+1]}$ when free energy is minimized.

We can go one step further than equation 4.20 and treat the hidden cause (v) as if it were data being generated by a higher hierarchical level, with slower dynamics (such that v appears not to change at the lower level). In doing so, we can chain together a hierarchy of equations:

$$
\begin{bmatrix} \vdots \\ \dot{\tilde{\mu}}_x^{(i)} - D\tilde{\mu}_x^{(i)} \\ \dot{\tilde{\mu}}_v^{(i)} - D\tilde{\mu}_v^{(i)} \\ \vdots \end{bmatrix} = \begin{bmatrix} \vdots \\ \nabla_{\tilde{\mu}_x^{(i)}} \tilde{g}^{(i)} \cdot \tilde{\Pi}_v^{(i-1)} \tilde{\varepsilon}_v^{(i-1)} - D \cdot \tilde{\Pi}_x^{(i)} \tilde{\varepsilon}_x^{(i)} + \nabla_{\tilde{\mu}_x^{(i)}} \tilde{f}^{(i)} \cdot \tilde{\Pi}_x^{(i)} \tilde{\varepsilon}_x^{(i)} \\ \nabla_{\tilde{\mu}_v^{(i)}} \tilde{g}^{(i)} \cdot \tilde{\Pi}_v^{(i-1)} \tilde{\varepsilon}_v^{(i-1)} + \nabla_{\tilde{\mu}_v^{(i)}} \tilde{f}^{(i)} \cdot \tilde{\Pi}_x^{(i)} \tilde{\varepsilon}_x^{(i)} - \tilde{\Pi}_v^{(i)} \tilde{\varepsilon}_v^{(i)} \\ \vdots \end{bmatrix}
$$

(4.21)

$$
\begin{bmatrix} \tilde{\varepsilon}_x^{(i)} \\ \tilde{\varepsilon}_v^{(i)} \end{bmatrix} = \begin{bmatrix} D\tilde{\mu}_x^{(i)} - f^{(i)}(\tilde{\mu}_x^{(i)}, \tilde{\mu}_v^{(i)}) \\ \tilde{\mu}_v^{(i)} - g^{(i+1)}(\tilde{\mu}_x^{(i+1)}, \tilde{\mu}_v^{(i+1)}) \end{bmatrix}
$$

$$
\tilde{\varepsilon}_v^{(0)} \triangleq \tilde{\varepsilon}_y
$$

Figure 4.6 graphically emphasizes the role of the hidden states (x) in linking together temporal derivatives within one hierarchical level and the role of the hidden causes (v) in linking hierarchical levels together. In this predictive coding scheme (Rao and Ballard 1999, Friston and Kiebel 2009), higher levels send descending predictions to lower levels, which compute errors in these predictions and pass these errors back up the hierarchy to update beliefs.

To complete our overview of predictive coding in the context of Active Inference, we need to incorporate action. Given that our aim is to minimize free energy and that the consequences of action are that we change our sensory data, we have the following:

$$
\dot{u} = -\nabla_u F
$$

(4.22)

$$
= -\nabla_u \tilde{y}(u) \cdot \tilde{\Pi}_y \tilde{\varepsilon}_y
$$

This equation says that we minimize free energy through action and that the only part of the free energy that depends directly on action is the lowest level of prediction error. In other words, action simply fulfills descending

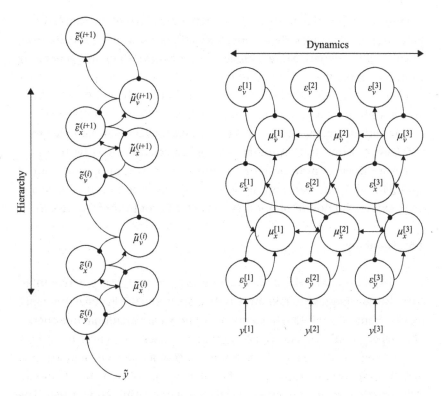

Figure 4.6
Message passing of generalized predictive coding schemes. *Left:* Computation of prediction errors from sensory data, showing how these may be propagated upward through a hierarchy. Higher levels send predictions to the lower levels that may be compared with sensory data to compute these errors. *Right:* A single layer of the hierarchy illustrates how neuronal populations representing different orders of generalized motion interact with one another.

predictions about data through minimizing the error between the predicted and observed sensory consequences of action. One way to think about this is as if we had equipped a predictive coding scheme with classical reflex arcs at the lowest level of the hierarchy (Adams, Shipp, and Friston 2013). In this setting, Active Inference is just predictive coding plus reflex arcs. From a neurobiological perspective, the idea is that sensory afferents enter the brain stem or spinal cord and synapse on motor neurons. Descending predictions of the sensory input are propagated from the cortex to the motor neurons, whose output depends on the difference between their cortical and sensory inputs.

From a computational perspective, a reflex arc is one of the simplest possible forms of controller; these correct deviations in predicted and observed proprioceptive signals. More complex motor behavior requires generating sequences of predictions and fulfilling them in order using reflex arcs. This mechanism sets active inference apart from other schemes for biological motor control, such as optimal control, which are not based on predictive coding and use inverse models and controllers that are more complex than reflex arcs (Friston 2011). Another peculiar characteristic of Active Inference is that it dispenses from notions of value or cost used in optimal control (and reinforcement learning); these are fully absorbed into the (generally more expressive) notion of priors (see chapter 10 for further discussion).

4.6   Summary

This chapter outlined the basic formal ideas that underwrite Active Inference. The key message to take away is that (approximate) Bayesian inference may be framed as minimizing a quantity known as variational free energy. This depends on a generative model that expresses our beliefs about how data are generated. We have looked at two forms of a generative model that may be employed depending on the inference problem at hand: specifically, whether we are interested in categorical or continuous variables. The free energy minimizing solution to either can be unpacked in terms of message passing between populations of neurons, including the generalized predictive coding schemes that follow from continuous models. Finally, we noted that free energy can be minimized not just by changing beliefs—such that they become consistent with data—but also by acting on the world to make data more consistent with beliefs. Over subsequent chapters, we will appeal to the formalisms introduced here and apply them to more concrete settings, providing an opportunity to explore the extensions of the broad concepts set out here.

5 Message Passing and Neurobiology

Basically there are two types of animals: animals, and animals that have no brains; they are called plants. They don't need a nervous system because they don't move actively, they don't pull up their roots and run in a forest fire! Anything that moves actively requires a nervous system; otherwise it would come to a quick death.

—Rodolfo Llinas

5.1 Introduction

In chapter 4, we saw the form that variational inference takes for two kinds of generative model. In this chapter, we focus on the *process theories* that arise from these inferential dynamics: theories that explain how the brain may implement variational inference. Central to this implementation of Bayesian belief updating is the notion of Bayesian message passing, which encompasses belief propagation and variational message passing, among other schemes. The idea subtending these schemes is that everything does not directly depend on everything else. Instead, each variable in a generative model depends on relatively few other variables. Similarly, the brain exhibits a sparse connectivity structure, wherein the activity of any neuron depends only on those neurons with which it shares synapses. This chapter focuses on the way we can map the sparse message passing associated with variational inference to the sparse connectivity structure of biological computation.

Let us take a step back from the technical material of chapter 4 and turn our attention to the process theories accompanying Active Inference. It is important to draw a distinction between a principle (i.e., the minimization of free energy) and a process theory about how this principle may be implemented in a certain kind of system, such as the brain. The latter lets us develop hypotheses that are answerable to empirical data. To address the ways in which Active Inference may manifest in the brain, we equate the message passing we saw at the end of chapter 4 with synaptic communication and the dynamics of gradient descent with neuronal activity. The dual aim of this chapter is to introduce readers with a technical background to the neurobiology of Active Inference and to highlight to biologists the relevance of theory to practical neuroscience. We emphasize that this chapter is not intended as the final word on process theories for Active Inference (Pezzulo, Rigoli, and Friston 2015, 2018; Friston and Buzsaki 2016; Friston and Herreros 2016; Friston, FitzGerald et al. 2017; Friston, Parr et al. 2017; Parr and Friston 2018b; Parr, Markovic et al. 2019); it is simply the interpretation that seems most consistent with currently available evidence. Nor is our aim to endorse a specific process theory but to illustrate how the ideas formulated in chapters 1–4 may be put to work in formulating hypotheses answerable to neurobiological measurements.

This chapter is organized as follows. First, in section 5.2, we consider the role of a cortical microcircuit. This is a functional unit comprising several neural populations connected to one another. The pattern of connectivity is replicated over many cortical regions. We highlight the relationship between this stereotyped circuit and the message passing architectures of figures 4.4 and 4.6—themselves recapitulated over hierarchical levels. In section 5.3 we move to effector systems and the formulation of motor control under Active Inference. This deals with the way in which the motor cortex tunes spinal and brain stem reflex arcs to generate purposeful behavior. Section 5.4 touches on ideas relating to subcortical structures like the thalamus and basal ganglia—which have important roles in planning and decision-making. We then consider, in section 5.5, modulation of synaptic efficacy, including the role of neurotransmitters in precision optimization and of plastic changes in learning. Finally, in section 5.6 we return to the theme of hierarchy and the relationship between decision-making and movement generation.

5.2   Microcircuits and Messages

In chapter 4, we saw that the belief-update equations mediating variational inference may be interpreted in terms of a (neuronal) network. The inference schemes presented—for continuous and categorical models—each give rise to a stereotyped circuitry whose structure is repeated in hierarchical generative models. Similarly, the architecture of the cerebral cortex has a stereotyped structure (Shipp 2007). The neocortex is divided into six layers (or laminae), numbered from superficial (close the brain's surface) to deep (closer to the subcortical white matter). Each layer is characterized by the presence of specific cell types and patterns of connectivity (Zeki and Shipp 1988, Felleman and Van Essen 1991, Callaway and Wiser 2009); this connectivity is summarized in the schematic of a single cortical column in figure 5.1.

A cortical column in one region of the brain connects to columns in other regions and to subcortical structures. Cortical regions are often depicted in a hierarchy that (loosely speaking) assigns those regions closest to sensory input or motor output to the lowest rungs of the hierarchy. As we move further away from these regions—for example, from primary to secondary visual cortex—we ascend the hierarchy. This notion of hierarchy is licensed by the laminar-specific connectivity structure illustrated in figure 5.1. Ascending projections (i.e., connections) from lower cortical areas or sensory (primary) thalamic nuclei tend to target the spiny stellate cells in layer IV. Lower cortical areas give rise to ascending connections from their superficial pyramidal cells (layers II and III). Descending projections from higher cortical areas target both superficial and deep layers of lower areas, with origins in the deep (layer VI) pyramidal cells. In addition, deep pyramidal cells (of Betz) in layer V project to various other targets, including subcortical nuclei—like the basal ganglia and secondary thalamic nuclei—and spinal motor neurons.

The middle schematic in figure 5.1 shows one possible mapping from the network for predictive coding (figure 4.4) to the laminar anatomy of the cortex (Friston, Parr, and de Vries 2017). This is a little complicated to interpret, but the key points are as follows. The ascending input to layer IV spiny stellate cells is associated with the prediction error for hidden causes ($\tilde{\varepsilon}_v^{(i)}$). The ascending output from layer III superficial pyramidal cells represents the same prediction error for the next layer of the hierarchy ($\tilde{\varepsilon}_v^{(i+1)}$). Descending input represents the prediction ($\tilde{g}^{(i+1)}$) from the higher level, while descending output is the prediction for the lower level ($\tilde{g}^{(i)}$). At the lowest level, we see descending predictions coming from layer V, consistent with the output to spinal motor neurons shown on the left. We will return to this in section 5.3. Recall from chapter 4 that the role of hidden causes is to link together hierarchical levels of a model that operates over multiple different timescales. This contrasts with the hidden states, whose role is

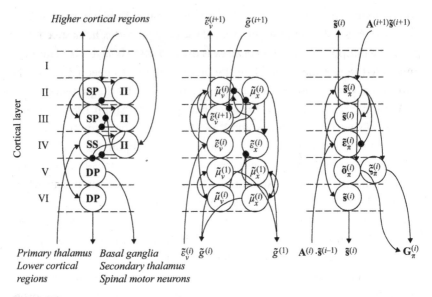

Figure 5.1

Canonical cortical microcircuits illustrating the relationship between inferential message passing and the architecture of the cerebral cortex. *Left:* Simplified schematic based on a synthesis of Miller 2003; Haeusler and Maass 2007; Shipp 2007, 2016; and Bastos et al. 2012 (refer to these papers for a summary of the neuroanatomical observations from which this synthesis is derived). Round arrowheads indicate inhibition; normal arrowheads indicate excitatory connections. The neural populations divide coarsely into superficial pyramidal (SP), deep pyramidal (DP), spiny stellate (SS), and inhibitory interneurons (II). *Middle:* Message passing that underwrites hierarchical predictive coding. *Right:* Message passing needed to solve a partially observable Markov decision process (POMDP).

in the dynamics at a specific timescale—consistent with their role in the intrinsic (within-column) connectivity in figure 5.1.

Asymmetry in message passing is important, as it offers empirical predictions about the difference between ascending and descending activity. One of these predictions is that we might expect these messages to be communicated by neural activity at different temporal frequencies. The reason for this is that the operations required to compute a prediction error from an expectation are nonlinear (Friston 2019b). This nonlinearity is due to the computation of predictions using nonlinear functions (g) that tend to increase the frequency of a signal (e.g., a doubling of frequency on squaring a sine wave). A prediction arising from this is that ascending

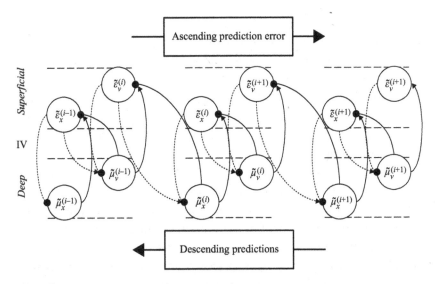

Figure 5.2
Simplified version of the predictive coding scheme shown in the middle of figure 5.1, unpacked to show the message passing between three cortical regions. Emphasis is on the asymmetry in message passing, with dotted lines showing ascending messages (prediction errors) and solid lines showing descending messages (predictions). Figure 5.1 is a finer-grained version of this schematic, including the intermediate neurons in the polysynaptic connections shown here. In this figure, coarse-graining the laminar specificity and dividing the cortex into Superficial and Deep relative to layer IV recovers a predictive coding scheme that will be familiar to many readers.

messages—originating from error units—may be measurable in higher frequency bands than descending messages—originating from expectation units (see figure 5.2). This is consistent with measured spectral asymmetries, wherein ascending connections are typically associated with gamma frequencies and descending connections with alpha or beta bands (Arnal and Giraud 2012; Bastos, Litvak et al. 2015).

The schematic on the right in figure 5.1 shows an interpretation of the message passing for a POMDP model as a cortical microcircuit. This has a structure similar to the predictive coding architecture, with expectations (s) represented in superficial and deep pyramidal cells and propagated up and down cortical hierarchies. In addition, error units (ε) in layer IV are in receipt of ascending signals. In contrast to predictive coding–style architectures, the messages passed between regions are mixtures of expectations as opposed to errors (Friston, Rosch et al. 2017; Parr, Markovic et al. 2019). However, the overall structure of minimizing an error (i.e., free energy gradient) by updating expectations is preserved. This message passing distinguishes between expectations conditioned on a policy (subscript π) and those averaged under policies. To translate from the former to the latter, we also need posterior beliefs about policies (π). We will return to this in section 5.4, but for now it is worth highlighting the consistency of this message passing with the targeting of superficial cortical layers by subcortical structures that could compute these beliefs.

In figure 5.1, note the absence of a one-to-one mapping between the architecture on the left and the message passing schemes in the middle and right. For example, there appears to be a discrepancy between the middle and left: the descending input in the column on the left arrives at layers II and IV, but that in the middle graphic targets layer III. This highlights that the connections implied by message passing schemes may not manifest as single synapses. The descending inhibitory connection targeting layer III could be mediated by the combination of an excitatory projection to layer IV inhibitory interneurons, and the projection of these interneurons to layer III. This disynaptic pathway resolves the apparent discrepancy between the two architectures.

In sections 5.3 and 5.4, we deal with the layer V neurons' role in movement and planning, corresponding to their spinal (or brain stem) and subcortical projections, respectively. In sections 5.5 and 5.6, we deal with the ways neural message passing is modulated over time and then return to the relationship between the microcircuits for categorical and continuous inference.

5.3   Motor Commands

The schematic on the left in figure 5.1 shows that layer V of the cortex projects to spinal pyramidal neurons and that this can be interpreted as a prediction (Adams, Shipp, and Friston 2013). This is unpacked in figure 5.3, which shows the spinal components of this circuit; we see a prediction based on the expectations encoded by Betz cells in layer V of the primary motor cortex. This is subtracted from the incoming proprioceptive input to the dorsal horn of the spinal cord, resulting in a proprioceptive prediction error. This error drives muscle activity that results in its own suppression—as proprioceptive data change to comply with predictions. The idea of a motor command as a prediction is central to Active Inference, as it highlights the

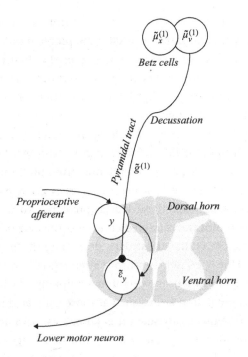

Figure 5.3
Neuroanatomy associated with Active Inference in modulating spinal motor reflexes. Starting from the Betz cells (upper motor neurons) in layer V of the motor cortex, the pyramidal tract carries predictions of proprioceptive input under the movement entailed by motor cortical expectations. The tract decussates (crosses over) and synapses—sometimes polysynaptically—on lower motor neurons in the ventral horn of the spinal cord. Subtracting the predictions from proprioceptive afferent signals arriving at the dorsal horn of the spinal cord results in an error that says how much muscle contraction would be required to meet the prediction. Lower motor neurons then cause this muscle contraction (or relaxation), ensuring that the resulting proprioceptive data match descending predictions.

duality of action and perception. The action part is the minimization of any discrepancy between predictions and sensory data by changing the data. This says that the only thing we should need to generate a movement is a prediction of the sensory consequences anticipated if that movement were to be executed. The fact that proprioceptive prediction errors can always be resolved by reflexes (as opposed to belief updating) offers a possible explanation for the paucity of granular cells in layer IV of the primary motor cortex (Shipp, Adams, and Friston 2013).

An important aspect of this sort of motor control is the notion of sensory attenuation (Brown et al. 2013). Consider the problem of initiating a new movement. To do this, we need to be able to predict that we are moving. However, until we are moving, the proprioceptive data to hand contradicts this hypothesis and could prompt its revision. Therefore, we need a way to preclude sensory data from updating our expectations, so that we can entertain the (initially false) belief that we are moving so that this belief can be realized through action (cf. ideomotor phenomena). The implication is that we need to be able to attend away from proprioceptive data by turning down their gain. Technically, this gain is given by the precision (inverse variance) with which these data are predicted. To attenuate this, we need descending motor tracts to predict not just the data but the precision of—or confidence placed in—those data, decreasing this precision to initiate movement. This is known as sensory attenuation and can be thought of as the complement to sensory attention, equipping us with the capacity to ignore certain prediction errors, such as those generated by saccadic eye movements (here sensory attenuation is known as saccadic suppression). Failures to attenuate are thought to be central to a range of neurological and psychiatric syndromes, including passivity phenomena (Pareés et al. 2014) and—at its most extreme—catatonic states in schizophrenia and the failure to initiate movements in Parkinson's disease.

5.4 Subcortical Structures

In addition to its projections to the spinal cord, cortical layer V targets several other structures. Among these is the striatum (Shipp 2007, Wall et al. 2013)—a structure deep within the cerebrum comprising the caudate nucleus and putamen. The striatum is the input nucleus of a complex network of structures known as the basal ganglia. Medium spiny neurons are the functional units of the striatum, taking input from the cortex and projecting to other nuclei of the basal ganglia. These divide into two types—those that express D1 dopamine receptors and those that express D2 receptors—where dopamine enhances activity of the former and attenuates it for the latter. The former cells are the origin of the direct pathway through the basal ganglia, connected by a single inhibitory synapse to the output nuclei (the internal globus pallidus and substantia nigra pars reticulata). The D2 cells give rise to the indirect pathway, a slightly more complex course with two inhibitory and one excitatory synapse. The striatum inhibits the external globus pallidus, which itself inhibits the subthalamic nucleus (STN). The STN projects to the basal ganglia outputs, meaning that the output nuclei are inhibited by the direct pathway and disinhibited by the indirect pathway. As these nuclei are themselves inhibitory, the net result of activating D1-expressing striatal neurons is disinhibition of behavior, which would be suppressed by D2-expressing neurons (Freeze et al. 2013).

Given that we have associated proprioceptive predictions with the projections to the spinal cord, which messages should we associate with the layer V projections to the striatum? Inspection of figure 5.1 offers a possible solution. Predicted outcomes (o) and the differences between preferred and predicted outcomes (ς) are shown in this layer, combining to compute the expected free energy (G) of a policy. Computing the last of these in the striatum is consistent with the notion that the basal ganglia are involved in planning—that is, evaluating alternative courses of action. Figure 5.4 shows one possible mapping of the message passing for policy evaluation onto the anatomy of the basal ganglia.

The key thing to draw from figure 5.4 is that, as described in chapter 4, posterior probabilities over policies (π)—shown here in the internal globus pallidus—are computed on the basis of their expected free energy. This pattern follows the direct pathway from layer V of the cortex through the striatum to the basal ganglia output nuclei. However, there are a few additional subtleties. In the hierarchical scheme shown on the left in figure 4.4, we see that expectations about states at higher levels can influence beliefs about policies at lower levels. Figure 5.4 shows this on the left, where the expected observations under high-level states are used to form empirical priors (E) that influence policy selection independently of the expected free energy. We will return to this in chapter 7, but the main idea is that we have beliefs about how we act in particular contexts. These prior expectations tend to bias policy evaluation when we find ourselves in the same context again—much like habit formation. In this sense, the influences of E and G can be seen as habitual and goal-directed drives, respectively. In reinforcement learning (Lee et al. 2014), these are sometimes referred to as "model-free" and "model-based" systems.[1] Associating these with the direct and indirect pathways of the basal ganglia has an interesting consequence: it implies that dopamine modulates the balance of the two. Remember that dopamine tends to promote the direct pathway and execution of specific policies (Moss and Bolam 2008)—presumably those associated with the lowest expected free energy. In contrast, low dopamine might be expected to favor context-sensitive priors in the indirect pathway, whose role is to suppress implausible policies in a given context. In a sense, striatal dopamine can be thought of as modulating the balance between inferring what to do and what not to do (Parr 2020).

The above is consistent with perturbations of the dopaminergic system; its depletion in severe Parkinson's disease causes akinesia—a failure to enact specific policies—while exogenous dopamine agonists promote impulsive behaviors (Frank 2005; Galea et al. 2012; Friston, Schwartenbeck et al. 2014). In addition, it is consistent with conceptual models of basal ganglia function. For example, Nambu (2004) suggests that the direct pathway mediates a fast and focused inhibition of the internal globus pallidus, followed by a broad and slow excitation, which causes excitation and inhibition of the targets of the basal ganglia, respectively. This is thought to ensure a "centre-surround" pattern that facilitates motor programs with a high specificity, which is consistent with the fast processes computing the expected free energy facilitating action and the broader contextualization of the slower pathway communicating empirical priors.

The final observation to make about figure 5.4 is that there are two levels of a cortical hierarchy (superscripted) contributing to the same basal ganglia circuit. This suggests temporally slower regions in targeting of indirect pathway neurons, but both fast and slow influences over the direct pathway. As we ascend cortical hierarchies, neurons tend to represent slower dynamics. For example, we may expect frontal cortical regions to represent longer timescales than parietal regions. This is consistent with the anatomical distribution of cortical inputs to the basal ganglia pathways (Wall et al. 2013). This temporal coarse graining in the indirect pathway is complemented by spatial coarseness, with the direct pathway medium spiny neurons exhibiting larger dendritic arbors (Gertler et al. 2008), enabling finer tuning. Therefore, the anatomy of figure 5.4 is endorsed by evidence from both clinical pathology (e.g., Parkinsonism) and cellular morphology.

In addition to the basal ganglia, many other important subcortical structures contribute to neuronal message passing. In the next section, we will discuss those from which neuromodulatory systems originate; we will conclude this section by briefly touching on the thalamus. We will not be able to do this highly complex structure full justice; however, we can outline some basic principles. The thalamus is often divided into primary and secondary nuclei. Figure 5.1 shows that primary thalamic nuclei can play the same role as lower cortical regions, in the sense that they target layer IV of the cortex and receive input from layer VI deep pyramidal cells (Thomson 2010, Olsen et al. 2012). An example is the lateral geniculate nucleus in the visual system, often thought of as a relay between the eye and the visual cortex. Like those neurons representing prediction errors, this receives both sensory information from the eye and backward projections from the cortex, which could be construed as predictions. Second order thalamic nuclei include the mediodorsal nucleus and the pulvinar, which interact with frontal and posterior cortices, respectively. These may have a role in predicting second order statistics (i.e., precision and variance) of sensory or higher order inputs and have been associated with figure-ground discrimination tasks (Kanai et al. 2015). Simplistically, this suggests that the division of the thalamus into primary and secondary nuclei may be a manifestation of the division into first and second order statistics.

5.5 Neuromodulation and Learning

Structural neuroanatomy is important, but it only gives us part of the picture of neural processing because the presence of a connection does not tell us much about the way it is used. As an example, consider the role of the substantia nigra shown in figure 5.4. The modulatory effect this has on striatal connectivity underwrites very different outputs from the basal ganglia, depending on the amount of dopamine released. Fast modulation of synaptic efficacy of this form can be contrasted with the slower but more persistent changes that occur with learning. In this section, we focus on these two ways in which synaptic efficacy can change.

Precision is an important concept in understanding neuromodulation (Feldman and Friston 2010). We touched on this in chapter 4 and in our discussion of sensory attenuation above, where we saw that it acts as a multiplicative weight on the prediction errors. More broadly, precision is a measure of confidence in a probability distribution. The relationship between the two is simple. If we have very precise beliefs about how data are generated from hidden states, then our beliefs about hidden states can be updated by observing data more than if we are not confident in those

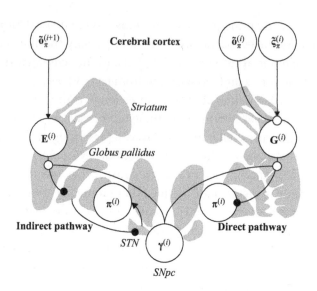

Figure 5.4
Direct and indirect pathways of the message passing for policy selection associated
with Active Inference through the basal ganglia using a POMDP generative model.
Pathways from the cerebral cortex culminate in estimation of policies. The direct
pathway (*right*) goes from cortex to striatum to internal globus pallidus. The indirect pathway (*left*) goes from cortex to striatum to external globus pallidus to subthalamic nucleus (STN) to internal globus pallidus. Both pathways exist bilaterally; in
addition, the substantia nigra pars compacta (SNpc) is shown modulating the balance between the two. (Note: This is a simplification of basal ganglia connectivity.)

beliefs. When belief updates manifest as changes in neuronal firing, a more
precise likelihood distribution manifests as an increased neural response
to a given sensory stimulus. This is essential for cognitive functions from
attention (Parr and Friston 2019a) to multisensory integration (Limanowski
and Friston 2019).

This perspective on synaptic gain control tells us something simple but
important. If precision is an attribute of some distribution in a generative
model, then there should be different precisions associated with different
distributions. This is intuitively sensible, as we can be more or less confident in the reliability of our sensations, in how things will dynamically
evolve, and even in how we might act (Parr and Friston 2017b). We have
seen the last of these in the context of dopaminergic signaling in the basal
ganglia. The greater the associated precision, the more confident we are
that our policies will minimize expected free energy.

If one of the roles of the dopaminergic system—originating in the substantia nigra pars compacta and the ventral tegmental area of the midbrain—is to signal confidence in what to do, then do other neuromodulatory systems play similar roles? Table 5.1 summarizes the evidence associating precisions with neuromodulatory systems—with the symbols sometimes used for these precisions. Specifically, the cholinergic system arising from the basal nucleus of Meynert appears to signal the precision of some likelihood distributions. The noradrenergic system, from the locus coeruleus, seems to play a role in signaling the precision of transitions over time. The serotonergic system seems less clear but may relate to the precision of prior preferences.

Why is it useful to be able to associate these precisions with neuromodulatory systems? The answer is threefold: it lets us explain observed biology, form hypotheses, and develop noninvasive methods to measure precision. We will highlight one example of each of these. First, regarding explanations of observed biology, empirical measurements of dopamine signals famously look like "reward prediction errors" (Schultz 1997)—with animals' dopamine increasing on receiving unexpected fruit juice or seeing a cue signaling imminent fruit juice. Active Inference offers an alternative explanation of these findings (Schwartenbeck, FitzGerald, Mathys, Dolan, and Friston 2015). Achieving a reward (or fulfilling our preferences) or encountering a cue indicative of a future reward enhances our confidence that we are pursuing a policy that minimizes expected free energy. This increase in confidence manifests as a spike in dopamine.

Second, regarding formation of hypotheses, an example concerns the decrease in cholinergic signaling associated with Lewy body dementia (Parr, Benrimoh et al. 2018)—a condition that leads to complex visual hallucinations. One plausible explanation for this is that accumulation of pathology in higher visual cortices prompts a mismatch between the predictions from these areas and the activity in primary visual cortices. Such a mismatch downgrades confidence in the associated likelihood distributions and causes loss of cholinergic signaling. The consequence of this loss of precision is a failure to update beliefs on the basis of sensory data, meaning perception loses the constraints afforded by sensation. This could explain the development of hallucinatory percepts in this condition.

Third, regarding noninvasive measurement of precision parameters, an example is the identification of computational phenotypes. There are a number of peripheral manifestations of central neurochemical activity,

Table 5.1

Putative roles of neurotransmitters in Active Inference

Neurotransmitter	Precision	Evidence
Acetylcholine	Likelihood (ζ)	• Presence of presynaptic receptors on thalamocortical afferents (Sahin et al. 1992, Lavine et al. 1997) • Modulation of gain of visually evoked responses (Gil et al. 1997, Disney et al. 2007) • Changes in effective connectivity with pharmacological manipulations (Moran et al. 2013) • Modeling of behavioral responses under pharmacological manipulation (Vossel et al. 2014, Marshall et al. 2016)
Noradrenaline	Transitions (ω)	• Maintenance of persistent prefrontal (delay-period) activity (requiring precise transition probabilities) depends on noradrenaline (Arnsten and Li 2005, Zhang et al. 2013) • Pupillary responses to surprising (i.e., imprecise) sequences (Nassar et al. 2012, Lavín et al. 2013, Liao et al. 2016, Krishnamurthy et al. 2017, Vincent et al. 2019) • Modeling of behavioral responses under pharmacological manipulation (Marshall et al. 2016)
Dopamine	Policies (γ)	• Expressed postsynaptically on striatal medium spiny neurons (Freund et al. 1984, Yager et al. 2015) • Computational fMRI reveals midbrain activity with changes in precision (Schwartenbeck, FitzGerald, Mathys, Dolan, and Friston 2015) • Modeling of behavioral responses under pharmacological manipulation (Marshall et al. 2016)
Serotonin	Preferences or interoceptive likelihood (χ)	• Receptors expressed on layer V pyramidal cells (Aghajanian and Marek 1999, Lambe et al. 2000, Elliott et al. 2018) in medial prefrontal cortex • Medial prefrontal cortical regions heavily implicated in interoceptive processing and autonomic regulation (Marek et al. 2013, Mukherjee et al. 2016)

Source: Parr and Friston 2018.

including the relationship between spontaneous blink rate and dopamine (Karson 1983) and between pupillary size and noradrenaline (Koss 1986). Recent work exploring the latter (Vincent et al. 2019) has demonstrated a relationship between the transition precision expected to be inferred by an ideal Bayesian observer and the dynamics of pupillary constriction and dilatation. The implication is that we could probe someone's implicit generative model (i.e., empirical prior beliefs) through peripheral measurements of this sort.

While fast changes in precision are important, this is a crude way of optimizing effective connectivity; it leads to an increase or decrease in the gain of a signal, but nothing more subtle. If we want to change the way the signal is interpreted, we need to rely on learning. We will return to this in detail in chapter 7. However, the basic idea is that we hold beliefs not just about states of the world but also about the fixed (or slowly varying) parameters that determine the dependencies between variables (Friston, FitzGerald et al. 2016). The substrate of these beliefs is the efficacy of synaptic connections between the neural populations representing time-varying variables (like hidden states or outcomes). When we observe an outcome that we believe was generated by a given state, we can update beliefs about the parameter connecting the two, reflecting an increase in the probability of them co-occurring in the future. In other words, we get a strengthening of the synapses between the two populations of neurons. The result is Hebb's famous edict (paraphrased): "Cells that fire together, wire together."

An important feature of figure 5.1 is that, in both predictive coding and marginal message passing schemes, the connections entering and leaving a cortical column relate to likelihood distributions. In contrast, transition probabilities and continuous dynamics depend on connections within a microcircuit. This suggests that learning dynamics should lead to changes in intrinsic connectivity, while learning observation models should modify extrinsic connectivity. Using techniques like dynamic causal modeling—which allow for evaluation of effective connectivity measures from neuroimaging data—it is possible to put these hypotheses to the test (Tsvetanov et al. 2016, Zhou et al. 2018). This highlights the role of process theories of this sort: they let us go beyond abstract theorizing to form specific testable hypotheses.

5.6 Continuous and Discrete Hierarchies

Finally, it is worth highlighting the move from continuous representations at low levels of a neural hierarchy to categorical variables at higher levels. The point is that the discrete and continuous message passing schemes we have considered likely coexist in the brain because we are able to hold beliefs of a categorical sort (e.g., in identifying what an object is or who a person is) in addition to being able to interface with continuously varying sensory receptors and effectors (e.g., muscle length or visual luminance contrast). This is reflected in neurophysiology, where some neurons are selective to specific stimuli and others vary in proportion to the intensity of a stimulus.

An interesting observation is that our interface with the world around us is in the continuous domain, the implication of which is that the lowest level of any hierarchy in the brain must be continuous. Having said this, we saw in figure 5.4 that policy selection in the basal ganglia may be framed as a discrete process, selecting between alternative movements. This tells us that we can think of movements as a composition of discrete trajectories into purposeful action. Where the lowest level might deal with the requisite changes in muscle length, descending input is based on decisions about which movement to make. From the perspective of a generative model, this means associating alternative (discrete) hypotheses about the world with the (continuous) dynamics entailed by those hypotheses. In chapter 8, we will return to the question of how to put these together from a computational perspective. Here, we simply note that the further we move from sensory receptors, the more we tend to find discretized representations in neural systems. Indeed, the very existence of classical receptive fields in neurophysiology could be interpreted as a probabilistic representation that the world is in some particular regime of a perceptual state-space—a state-space that is tiled by receptive fields and consequently partitioned into lots of little categories. Figure 5.5 brings together these schemes and acts as a summary of the ideas set out in this chapter.

5.7 Summary

This chapter has sought to outline the points of connection between the message passing schemes implied by the generative models of chapter 4 and the neurobiology of inference, action, and planning. What do we gain by

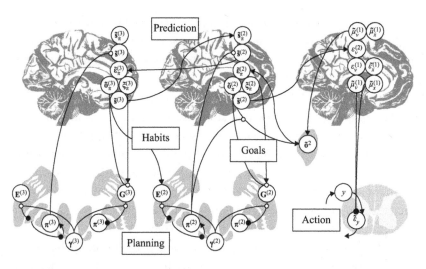

Figure 5.5
Anatomy of inference (based on Friston, Parr, and de Vries 2017) connects schematics
from figures 5.1–5.4, providing a summary of the ideas in this chapter. Two hierarchical loops through the cortex and basal ganglia highlight the distinction between
habits—based on input from higher levels—and the more context-sensitive, goal-
directed (explorative and exploitative) behavior resulting from expected free energy
minimization. Note the influence of inferences about policies on **s**, implementing
the Bayesian model averaging under policies referred to in the main text. This projection from basal ganglia to cortex may be mediated by intermediate structures, such as
the thalamus. On the right, the categorical POMDP-based messages are relayed into
a continuous predictive-coding network, involved in generation of action. Each categorical state is associated with an alternative prediction of continuous variables and
contributes to a prediction error. The message in the opposite direction computes the
posterior probability of the associated categorical outcome (**o**), which depends on priors based on the policy-dependent outcome (**o**$_\pi$), beliefs about the policy (**π**), and the
likelihood of the continuous trajectory that may be computed from posterior expectations (μ) and variances (not shown) at the continuous level. More connections could
be included here; for instance, in addition to habits (**E**), the selection of goals (**C** from
chapter 4) is itself likely to depend on higher hierarchical levels, leading to hierarchical
control of motivation (see Pezzulo, Rigoli, and Friston 2018 for details).

relating message passing to neuronal communication? It allows us to make empirical predictions based on the generative model that we hypothesize the brain is inverting. This may take the form of an evoked response—the change in potential that is measurable at the scalp on presenting the brain with a sensory stimulus—whose time course will depend on the amount of belief updating induced by that stimulus. Alternatively, computational neuroimaging methods can be used to associate simulated inferences with those brain regions exhibiting similar temporal dynamics (Schwartenbeck, FitzGerald, Mathys, Dolan, and Friston 2015). Making this association is important in understanding pathology—and therapeutics—for computational (i.e., neurological and psychiatric) disorders, allowing for expression of functional pathologies in terms of their biology.

Finally, it is worth acknowledging that much of the brain has been conspicuously absent in this chapter—partly for reasons of space but also because neuroscience is a work in progress. There are many opportunities to extend (or even replace) the account given in this chapter. To some degree we can extrapolate from what we have seen here. For example, parts of the amygdala are cytoarchitecturally equivalent to basal ganglia nuclei. Does this mean there is a class of policies evaluated by the amygdala? Could this structure be to autonomic policies what the basal ganglia are to those in the skeletomotor domain? Might other structures (like the pulvinar) play similar roles for other (e.g., mental) classes of policy? How should we understand cortical architectures that differ from the six-layered structure in figure 5.1? The cerebellum and the hippocampal formation each exhibit distinct but stereotyped microcircuitry (Wesson and Wilson 2011). Should we see these as anatomical rearrangements of the same Bayesian message passing schemes, or do they deal with different aspects of a generative model (Pezzulo, Kemere, and van der Meer 2017; Stoianov et al. 2020)? We raise these questions not to offer any answers but to highlight some of the exciting avenues of future research in theoretical neurobiology. Active Inference and its associated process theories offer a rigorous formal and conceptual framework in which to address these questions.

6 A Recipe for Designing Active Inference Models

Give me six hours to chop down a tree and I will spend the first four sharpening the axe.

—Abraham Lincoln

6.1 Introduction

This chapter provides a four-step recipe to construct an Active Inference model, discussing the most important design choices one has to make to realize a model and providing some guidelines for those choices. It serves as an introduction to the second part of the book, which will illustrate several specific computational models using Active Inference and their applications in a variety of cognitive domains.

As Active Inference is a normative approach, it tries to explain as much as possible about behavior, cognitive, and neural processes from first principles. Consistently, the design philosophy of Active Inference is top-down. Unlike many other approaches to computational neuroscience, the challenge is not to emulate a brain, piece by piece, but to find the *generative model* that describes the problem the brain is trying to solve. Once the problem is appropriately formalized in terms of a generative model, the solution to the problem emerges under Active Inference—with accompanying predictions about brains and minds. In other words, the generative model provides a complete description of a system of interest. The resulting behavior, inference, and neural dynamics can all be derived from a model by minimizing free energy.

The *generative modeling* approach is used in several disciplines for the realization of cognitive models, statistical modeling, experimental data analysis, and machine learning (Hinton 2007b; Lee and Wagenmakers 2014; Pezzulo, Rigoli, and Friston 2015; Allen et al. 2019; Foster 2019). Here, we are primarily interested in designing generative models that engender cognitive processes of interest. We have seen this design methodology in previous chapters. For example, using a generative model for predictive coding, perception was cast as an inference about the most likely cause of sensations; using a generative model that evolves in discrete time, planning was cast as an inference about the most likely course of action. Depending on the problem of interest (e.g., planning during spatial navigation or planning saccades during visual search), one can adapt the form of these generative models to equip them with different structures (e.g., shallow or hierarchical) and variables (e.g., beliefs about allocentric or egocentric spatial locations). Importantly, Active Inference may take on many different guises under different assumptions about the form of the generative model being optimized. For example, assumptions about models that evolve in discrete or continuous time influence the form of the message passing (see chapter 4). This implies that the choice of a generative model corresponds to specific predictions about both behavior and neurobiology.

This flexibility is useful as it allows us to use the same language to describe processes in multiple domains. However, it can also be confusing from a practical perspective, as there are a number of choices that must be made to find the appropriate level of description for the system of interest. In the second part of this book, we will try to resolve this confusion through a series of illustrative examples of Active Inference in silico. This chapter introduces a general recipe for the design of Active Inference models, highlighting some of the key design choices, distinctions, and dichotomies that will appear in the numerical analysis of computational models described in subsequent chapters.

6.2   Designing an Active Inference Model: A Recipe in Four Steps

Designing an Active Inference model requires four foundational steps, each resolving a specific design question:

1. **Which system are we modeling?** The first choice to make is always the system of interest. This may not be as simple as it seems; it rests on the identification of the boundaries (i.e., Markov blanket) of that system. What counts as an Active Inference agent (generative model), what

counts as the external environment (generative process), and what is the interface (sensory data and actions) between them?

2. **What is the most appropriate form for the generative model?** The first of the next three practical challenges is deciding whether it is appropriate to think of a process more in terms of categorical (discrete) inferences or continuous inferences, motivating the choice between discrete or continuous-time implementations (or a hybrid) of Active Inference. Then we need to select the most appropriate hierarchical depth, motivating the choice between shallow versus deep models. Finally, we need to consider whether it is necessary to endow generative models with temporal depth and the ability to predict action-contingent observations to support planning.

3. **How to set up the generative model?** What are the generative model's most appropriate variables and priors? Which parts are fixed and what must be learned? We emphasize the importance of choosing the right sort of variables and prior beliefs; furthermore, we emphasize a separation in timescales between the (faster) update of state variables that occurs during inference and the (slower) update of model parameters that occurs during learning.

4. **How to set up the generative process?** What are the elements of the generative process (and how do they differ from the generative model)?

These four steps (in most cases) suffice to design an Active Inference model. Once completed, the behavior of the system is determined by the standard schemes of Active Inference: the descent of the active and internal states on the free energy functional associated with the model. From a more practical perspective, once one has specified the generative model and generative process, one can use standard Active Inference software routines to obtain numerical results, as well as to perform data visualization, analysis, and fitting (e.g., model-based data analysis). In what follows, we will review the four design choices in order.

6.3 What System Are We Modeling?

A useful first step in applying the formalism of Active Inference is to identify the boundaries of the system of interest because we are interested in characterizing the interaction between what is *internal* to a system and the *external world* via sensory receptors and effectors (e.g., muscles or glands). As discussed in chapter 3, a formal way to characterize the distinction between internal states of a system and external variables (and intermediate variables that mediate their interactions) is in terms of a Markov blanket (Pearl 1988). To reiterate the argument, a Markov blanket may be subdivided into two sorts of variables (Friston 2013): those that mediate the influence of the external world on internal states of the system of interest (i.e., sensory states) and those that mediate the influence of internal states of the system of interest on the external world (i.e., active states). See figure 6.1.

Importantly, there are many ways in which a boundary between internal and external may be defined. In most of the simulations we will discuss in the second part of this book, there will be a (Markov blanket) separation between an agent (roughly, a living organism) and its environment. This corresponds to the usual setup of cognitive models, where an agent implements cognitive processes such as perception and action selection on the basis of its internal (e.g., brain) states and is provided with sensors and effectors.

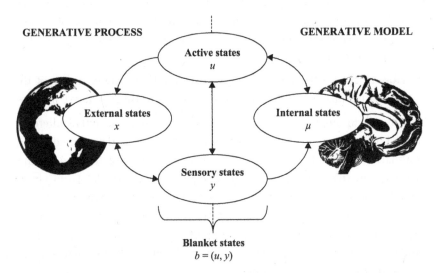

Figure 6.1
Action-perception loop between an adaptive system (here, the brain) and the environment, along with the Markov blanket (composed of active states and sensory states) that mediates their interaction. The figure implies that the adaptive system only affects the environment by performing actions (via active states) and that the environment only affects the adaptive system by producing observations (via sensory states). The figure exemplifies the distinction between the adaptive system's generative model and the (external) generative process that produces its observations.

However, this is not the only possibility. From the perspective of neurobiology, we could draw a Markov blanket around a single neuron, around the brain, or around the entire body. In the first case, sensory states include postsynaptic receptor occupancies, and active states include the rate at which vesicles containing neurotransmitters fuse with the presynaptic membrane. The internal states of the neuron (e.g., membrane potentials, calcium concentrations) can then be thought of as inferring the causes of its sensory states according to some (implicit) generative model (Palacios, Isomura et al. 2019). This setup treats the external states (that are being modeled) as including the neuronal network in which our neuron participates. This is very different from the inference taking place when we assume our entire network is internal to the Markov blanket. For example, if we take a system whose sensory states are the photoreceptors in the retina and whose active states are the oculomotor muscles, the inferences performed by the internal states are about things outside the brain. This speaks to the importance of scale, as the internal states of this Markov blanket include the internal states from the perspective of a single neuron. The latter internal states appear to make inferences about things within the brain when the Markov blanket is drawn around a single neuron but not when the blanket is drawn around the nervous system.

The above is particularly relevant when dealing with embodied or extended perspectives on cognition (Clark and Chalmers 1998; Barsalou 2008; Pezzulo, Lw et al. 2011). For example, if we draw the blanket around the nervous system, the rest of the body becomes an external state, about which we must make inferences from interoceptive sensory states (Allen et al. 2019). Alternatively, we could draw our blanket around the entire organism. This would make it look as if organs other than the brain were making inferences about their environment. For example, depression of the skin in response to an external pressure could be framed as an inference about the source of the external pressure. The extended cognition perspective takes this further and says that objects external to the body may be incorporated into the Markov blanket (e.g., the use of a calculator to assist in inference implies that the calculator is part of the internal state-space of the inferring system). Finally, we could have multiple Markov blankets, nested within one another (e.g., brains, organisms, communities).

In sum, defining the Markov blanket ensures we know what is being inferred (external states) and what is doing the inferring. Indeed, minimization of free energy with respect to a generative model only involves the internal and active states of a system: these only see the sensory states, so they can only infer the external state of the world vicariously.

6.4   What Is the Most Appropriate Form for the Generative Model?

Once we have decided on the internal states of a system and the states that mediate their interaction with the world outside, we need to specify the *generative model* that explains how external states influence sensory states.

As discussed in previous chapters, Active Inference can operate on different kinds of generative models. Therefore, we need to specify the most appropriate form of the generative model for the problem at hand. This implies making three main design choices. The first is a choice between models that include continuous or discrete variables (or both). The second is a choice between shallow models, in which inference operates on a single timescale (i.e., all variables evolve at the same timescale), and hierarchical or deep models, in which inference operates on multiple timescales (i.e., different variables evolve at different timescales). The third is a choice between models that only consider present observations versus models having some temporal depth, which consider the consequences of actions or plans.

6.4.1   Discrete or Continuous Variables (or Both)?

The first design choice is to consider whether generative models that use discrete or continuous variables are more appropriate. The former include object identities, alternative action plans, and discretized representations of continuous variables. These are modeled through expressing the probability—at each time step—of one variable transitioning into another type. The latter include things like position, velocity, muscle length, and luminance and require a generative model expressed in terms of rates of change.

Computationally, the distinction between the two may not be clear-cut because a continuous variable may be discretized, and a discrete variable may be expressed through continuous variables. However, this distinction is important conceptually, as it underlies specific hypotheses about the time course (discrete or continuous) of the cognitive processes of interest.[1] In most current implementations of Active Inference, high-level decision processes, such as the choice between alternative courses of actions, are modeled using discrete variables, whereas more fine-grained perception and action dynamics are implemented using continuous variables; we will provide examples of both in chapters 7 and 8, respectively.

Furthermore, the choice between discrete and continuous variables is relevant for neurobiology. While each style of modeling appeals to free energy minimization, the message passing these imply take different forms. To the extent that one considers message passing relevant for a process theory (see chapter 5), this implies that the neural dynamics that realize this minimization are different under each sort of model. Continuous schemes underwrite predictive coding—a theory of neural processing that relies on top-down predictions corrected by bottom-up prediction errors. However, the analogous process theories for discrete inferences involve messages of a different form. Finally, the two types of model may be combined such that discrete states are associated with continuous variables. This means we can specify a generative model wherein a discrete state (e.g., object identity) generates some pattern of continuous variables (e.g., luminance). We will discuss an example of a hybrid or mixed generative model that includes both discrete and continuous variables in chapter 8.

6.4.2   Timescales of Inference: Shallow versus Hierarchical Models

The second design choice concerns the timescales of Active Inference. One can select either (shallow) generative models, in which all the variables evolve at the same timescale, or (hierarchical or deep) models, which include variables that evolve at different timescales: slower for higher levels and faster for lower levels.

While many simple cognitive models only require shallow models, these are not sufficient when there is a clear separation of timescales between different aspects of a cognitive process of interest. One example of this is in language processing, in which short sequences of phonemes are contextualized by the word that is spoken and short sequences of words are contextualized by the current sentence. Crucially, the duration of the word transcends that of any one phoneme in the sequence and the duration of the sentence transcends that of any one word in the sequence. Hence, to model language processing, one can consider a hierarchical model in which sentences, words, and phonemes appear at different (higher to lower) hierarchical levels and evolve over (slower to faster) timescales that are approximately independent of one another. This is only an approximate separation, as levels must influence each other (e.g., the sentence influences the next words in the sequence; the word influences the next phonemes in the sequence).

However, this does not mean we need to attempt to model the entire brain to develop meaningful simulations of a single level. For example, if we wanted to focus on word processing, we could address some aspects without having to deal with phoneme processing. This means we can treat input from parts of the brain drawing inferences about phonemes as providing observations from the perspective of word-processing areas. Phrasing this in terms of a Markov blanket, this typically means we treat the inferences performed by lower levels of a model as part of the sensory states of the blanket. This means we can summarize the inferences performed at the timescale of interest without having to specify the details of lower-level (faster) inferential processes—and this hierarchical factorization entails great computational benefits.

Another example is in the domain of intentional action selection, where the same goal (enter your apartment) can be active for an extended period of time and contextualizes a series of subgoals and actions (find keys, open door, enter) that are resolved at a much faster timescale. This separation of timescales, whether in the continuous or discrete domain, demands a hierarchical (deep) generative model. In neuroscience, one can assume that cortical hierarchies embed this sort of temporal separation of timescales, with slowly evolving states at higher levels and rapidly evolving states at lower levels, and that this recapitulates environmental dynamics, which also evolve at multiple timescales (e.g., during perceptual tasks like speech recognition or reading). In psychology, this sort of model is useful in reproducing hierarchical goal processing (Pezzulo, Rigoli, and Friston 2018) and working memory tasks (Parr and Friston 2017c) of the sort that rely on delay-period activity (Funahashi et al. 1989).

6.4.3   Temporal Depth of Inference and Planning

The third design choice concerns the temporal depth of inference. It is important to draw a distinction between two kinds of generative model: the first have *temporal depth* and represent explicitly the consequences of actions or action sequences (policies or plans), whereas the second lack temporal depth and only consider present but not future observations. These two kinds of model are exemplified in figure 4.3: the dynamic POMDP at the top and the continuous-time model at the bottom.[2] The key difference between these two models is not that they use discrete or continuous variables, respectively, but that only the former (temporally deep) model endows creatures with the ability to plan ahead and select among possible futures.

Imagine a rodent who plans a route to a known food location in a maze. Doing this benefits from a temporally deep model, loosely equivalent to a spatial or cognitive map (Tolman 1948), which encodes contingencies between present and future locations conditioned on actions (e.g., the future location after turning right or left). The animal can use the temporally deep model to counterfactually consider multiple courses of action (e.g., series of right and left turns) and select the one expected to reach the food location.

Why is a temporally deep model required for planning? In Active Inference, planning is realized by calculating the expected free energy associated with different actions or policies and then selecting the policy that is associated with the lowest expected free energy. Expected free energy is not just a function of present observations (like variational free energy) but also a functional of future observations. The latter cannot be observed (by definition) but only predicted using a temporally deep model, which describes the ways in which actions produce future observations.

When designing an Active Inference agent it is useful to consider whether it should have planning and future-oriented capacities—and, in this case, to select a temporally deep model. Furthermore, it is useful to consider planning depth—that is, how far in the future the planning process can look. Finally, one can design generative models that are both hierarchical and temporally deep, wherein planning proceeds at multiple timescales—faster at lower levels, and slower at higher levels.[3] The decision whether to model alternative futures, contingent on policy selection, is largely tied up with the choice between discrete and continuous models because the idea of selecting between alternative futures, defined by sequences of actions, is more simply articulated using discrete-time models.

6.5 How to Set Up the Generative Model?

When we have specified our system of interest and identified the relevant forms of the generative model (e.g., continuous or discrete representation, shallow versus hierarchical structure), our next challenges are to specify the specific variables to include in the generative model and decide which of these variables remain fixed or change as an effect of learning.

6.5.1 Setting Up the Variables of the Generative Model

The variables of generative models can be either predefined or learned from data. For illustrative purposes, most models that we discuss in this book use predefined variables. When designing these models, in practice, the main challenge is deciding which hidden states, observations, and actions are most appropriate for the problem at hand. For example, the perceptual model able to distinguish frogs from apples in chapter 2 only included two hidden states (frogs, apples) and two observations (jumps, does not jump). A more sophisticated model could include additional observations (e.g., red, green) as well as actions such as touching, which produce differential sensory effects (jump or no jump) in the presence of a frog or an apple.

Figure 6.2 schematically illustrates a generative model for the concept of a jumping frog. The concept is cast as a hierarchical model, where

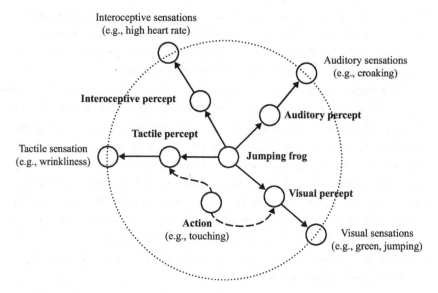

Figure 6.2
(Hierarchical) generative model for the concept of a jumping frog uses a simplified notation compared to chapter 4: nodes within the dotted circle correspond to hidden states, whereas nodes at the periphery correspond to sensory observations. Beliefs about hidden states, following inversion of the model, correspond to percepts that may be tied to a sensory modality (e.g., visual percept) or may be amodal (e.g., the jumping frog). Action contingencies are represented as dashed lines. Horizontal dependencies between hidden states in different modalities, as well as temporal dependencies between hidden states (as we saw in the dynamical generative models of chapter 4), are ignored for the sake of simplicity.

Box 6.1

Varieties of sensory modalities: Exteroceptive, proprioceptive, and interoceptive

In Active Inference, a conceptual distinction is often made between three kinds of sensory modalities: exteroceptive (e.g., vision and audition), proprioceptive (e.g., the sense of joint and limb positions), and interoceptive (e.g., the sense of the internal organs of the body, such as heart and stomach). In multimodal generative models, one can often factorize parts of the model that relate to different modalities; this permits representing that (for example) saccadic movements have visual but not auditory consequences.

Importantly, the same principles of Active Inference operate across all the modalities. For example, in the same way visual processing can be described as the inference about (hidden variables about) a perceptual scene, interoceptive processing can be described as the inference about (hidden variables that report) the internal state of the body. Furthermore, motor actions that change the perceptual scene and internally directed actions that change the interoceptive state can be described in a similar way. The former engages spinal reflexes that fulfill proprioceptive predictions, whereas the latter engages autonomic reflexes that fulfill interoceptive predictions. Such interoceptive processing supports allostasis and adaptive regulation, and its dysfunctions can have psychopathological consequences (Pezzulo 2013, Seth 2013, Pezzulo and Levin 2015, Seth and Friston 2016, Allen et al. 2019).

a single (multimodal or supramodal) hidden state at the center of the figure unfolds in a cascade of (unimodal) hidden states corresponding to percepts in different modalities (exteroceptive, proprioceptive, and interoceptive; see box 6.1) and ultimately causing sensations in the same modalities. This arrangement corresponds to casting the jumping frog concept as the *common cause* of multiple *sensory consequences* (e.g., something green and jumping in the visual domain; a croaking sound in the auditory domain), some of which can be action-contingent (e.g., the sight of something jumping may increase on touching it). The inversion of the generative model corresponds to a perceptual inference (e.g., the presence of a jumping frog) from its observed sensory consequences (e.g., the sight of something green and jumpy), and it integrates information across multiple modalities.

Once these variables of interest have been established, the next exercise is to write down the full generative model. One example is the simple generative model for frogs and apples in figure 2.1, which is fully specified by prior beliefs about hidden states and a (likelihood) mapping between hidden states and observations and whose numerical values can be either specified by hand or learned from data (see 6.5.2).

Beyond this simple example, the elements that need to be specified are fully determined by the form of the selected generative model. For example, the model for discrete-time POMDP shown in figure 4.3 (top) requires specifying the **A**, **B**, **C**, **D**, and **E** matrices; continuous schemes use analogous (although less alphabetical) elements, which will be dealt with in chapter 8. But even in these more complex cases, the exercise is not so dissimilar from above: namely, specifying prior beliefs about the variables of interest (e.g., in discrete-time implementations, about hidden states at the first time step in the **D**-vector and about observations in the **C**-matrix) and their probabilistic mappings (e.g., likelihood mapping between hidden states and observations in the **A**-matrix). However, in some cases, it is useful to think about *factorizations* of the state-space of the generative model, which avoids considering every possible combination of variables if some are unnecessary. In chapter 7, we will discuss a biologically plausible example of factorization that occurs in perceptual processing between "what" and "where" streams (Ungerleider and Haxby 1994)—namely, between variables that represent object identities and locations, respectively, which can be treated independently in the model (hence simplifying it) as they are often invariant to one another.

Deciding which variables are of interest and the ways they are related or factorized in the model is often the most challenging—but also the most creative—part of model design. It is an exercise of translating our cognitive hypotheses into a mathematical form that supports Active Inference. How should we select the "right" variables? Ultimately, this is a question of specifying plausible alternatives and picking those that have the lowest free energy (cf. Bayesian model comparison). However, a practically useful perspective for most studies is that the generative model should be as similar as possible to how we believe data are generated. When appealing to Active Inference in the setting of cognitive psychology, this often means thinking about how experimental psychologists would go about generating the stimuli they present to their experimental participants. On formalizing these processes in terms of the requisite probability distributions, we arrive at a generative model whose free energy minimizing dynamics naturally lead to performance of the task in question.

Here, we can draw an analogy with most Bayesian (or ideal observer) models of perception, in which the models are designed to mimic (to a

Box 6.2

Priors and empirical behavior

Another perspective on the issue of selecting priors draws from a set of results known as the *complete class theorems* (Wald 1947, Daunizeau et al. 2010), which state that any statistical decision procedure (i.e., behavior) may be framed as Bayes optimal under the right set of prior beliefs. This means that if we are interested in explaining empirical behavior, our challenge is to identify the generative model (comprising prior beliefs) that would reproduce that behavior as simply as possible. In short, priors are a statement of a hypothesis about the system in question. If other prior beliefs would be plausible, this offers an opportunity to put this to empirical data through Bayesian model comparison. This also has implications for computational phenotyping in clinical populations. That there will always be a set of prior beliefs that render behavior Bayes optimal implies the key question—in understanding the computational deficits that give rise to psychiatric or neurological syndromes—is what these priors are. This idea is slightly counterintuitive at first. However, the complete class theorem means that asking whether a behavior is (Bayes) optimal is meaningless. The important question is, What are the prior beliefs that would make this optimal? In chapter 9, we will see how an appeal to free energy minimization based on our own beliefs as scientists offers a way to answer this question.

large extent) the structure of the task at hand, as in the example of recognizing a frog or an apple (chapter 2). This idea is sometimes equated with the *good regulator theorem* (Conant and Ashby 1970), which says that to regulate an environment effectively, a creature (whether biological or synthetic) must be a good model of that system. From the perspective of eco-niche construction, this is sometimes phrased in terms of the (statistical) fitness (Bruineberg et al. 2018) of a creature's model to its environment (and vice versa). However, this does not mean that an agent's generative model has to be identical to the generative process that actually generates data. For most practical applications, it can be simplified or different. We will return to this point later in this chapter (6.6).

6.5.2 Which Parts of the Generative Model Are Fixed, and What Is Learned?

Another design choice is deciding which parts of the generative model are fixed and which ones are updated over time as an effect of learning. In principle, Active Inference allows every part of the model—and even its structure—to be updated (or learned) over time. This renders learning a design choice rather than something mandatory. In keeping with this, we will cover examples of Active Inference models that are completely designed by hand and examples in which some parts of the model (e.g., transition probabilities) remain fixed while others (e.g., likelihoods) are updated over time.

In Active Inference, learning is cast as an aspect of inference, as a free energy minimizing process. So far, we have described inference in terms of an *update of beliefs about states* of the generative model. In much the same way, we can describe learning as an *update of beliefs about parameters* of the generative model. For this, the generative model has to be endowed with *prior beliefs about parameters* of the distributions to be learned, where the specific parameters depend on probability distribution associated with each variable (e.g., *mean* and *variance* for a Gaussian distribution). These prior values are updated to form posterior beliefs whenever new data are encountered. As we will discuss in chapter 7, the algorithmic form of this update is the same as the update of state variables.

The fact that both inference and learning use the same kind of Bayesian belief updates may seem confusing during model design—partly because deciding what should be modeled as a state or a parameter is not always straightforward. However, when it comes to cognitive models, there is a clear difference between inference and learning. Inference describes (fast) changes of our beliefs about model states—for example, how we update our belief that there is an apple in front of us after observing something red. Learning describes (slow) changes of our beliefs about model parameters—for example, how we update our likelihood distribution to increase the value of the apples-red mapping after observing several occurrences of red apples. Beliefs about parameters typically vary much more slowly than those about states, and they may only be updated after states have been inferred. From a neurobiological perspective, it is appealing to map inference to neuronal dynamics and learning to synaptic plasticity. Furthermore, as we will discuss in chapter 7, holding probabilistic beliefs about model parameters induces novelty-seeking behaviors so that creatures may select the best data to learn the causal structure of their worlds. This suggests that endowing Active Inference models with the ability to learn their *parameters* (or even their *structure*; see chapter 7) is an effective way to study the behavioral dynamics of active learning and curiosity-based exploration.

Before concluding this section, it is worth noting that in this book we exemplify rather simple generative models that are defined using *tabular methods* (e.g., with explicit matrices for priors and likelihoods) and that operate in small state-spaces. In comparison, much more sophisticated kinds of generative models—and associated learning schemes—are being developed in fields like machine learning, deep learning, and robotics, such as, for example, variational autoencoders (Kingma and Welling 2014), generative adversarial networks (Goodfellow et al. 2014), recursive cortical networks (George et al. 2017), and world models (Ha and Schmidhuber 2018). In principle, one could borrow any of these methods (and many others) to implement one or more parts of Active Inference models (e.g., likelihood or transition models). By leveraging the most up-to-date machine learning methods, it would be possible to scale up Active Inference to increasingly more challenging domains and applications; see, for example, Ueltzhöffer (2018) and Millidge (2019).

However, there are some important points to consider when designing Active Inference models that use sophisticated machine learning models, especially if one is interested in cognitive and neurobiological implications. One appeal of Active Inference is that it offers an integrative perspective on cognitive functions by assuming that (for example) perceptual inference, action planning, and learning all stem from the same free energy minimization process. This integrative power would be lost if (for example) one juxtaposed generative models that operate or learn independently from one another. Furthermore, the aforementioned machine learning methods correspond to process models that are distinct from Active Inference and have different cognitive and neurobiological interpretations. Finally, when using machine learning methods, some of the design choices discussed here (e.g., about the choice of model variables) may be skipped, as they are emergent properties of learning; however, they may be replaced by different design choices, about (for example) number of layers, parameters, and learning rates of a deep neural net. These design choices potentially have relevant cognitive and neurobiological implications, which are beyond the scope of what we address here.

6.6 Setting Up the Generative Process

In Active Inference, the *generative process* describes the dynamics of the world external to the Active Inference agent, which corresponds to the process that determines the agent's observations (see figure 6.1). It may seem bizarre to have postponed defining the generative process until after describing the agent's generative model. After all, a modeler would have some task (and generative process) in mind from the beginning, so it would make perfect sense to revert this order and design the generative process before the generative model, especially in applications where the generative model has to be learned during situated interactions, as in gamelike or robotic settings (Ueltzhöffer 2018, Millidge 2019, Sancaktar et al. 2020).

The reason we postponed the design of the generative process is that, in many practical applications discussed in this book, we simply assume that the dynamics of the generative process are the same as, or very similar to, the generative model. In other words, we generally assume that the agent's generative model closely mimics the process that generates its observations. This is not the same as saying that the agent has perfect knowledge of the environment. Indeed, even if the agent knows the process that generates its observations, it may be uncertain about (for example) its initial state in the process, as was the case in the apple versus frog example. In the language of discrete-time Active Inference, one could design a model in which both the generative model and the generative process are characterized by the same A-matrix but in which the agent's belief about its initial state (D-vector), which is part of its generative model, is different from—or even inconsistent with—the true initial state of the generative process. One subtle thing to notice is that even if both the generative model and the generative process are characterized by the same A- and B-matrices, their semantics are different. The A-matrix of the generative process is an objective property of the environment (sometimes called a *measurement distribution* in Bayesian models), whereas the A-matrix of the generative model encodes an agent's subjective belief (called a *likelihood function* in Bayesian models).

Of course, except in the simplest cases, it is not mandatory that the generative model and generative process are the same. In practical implementations of Active Inference, one can always specify the generative process separately from the generative model, either using equations that differ from those of the generative model or using other methods, such as game simulators, which take actions as inputs and provide observations as outputs (Cullen et al. 2018), thereby following the usual action-perception loop implied by the Markov blanket of figure 6.1.

There are some philosophical implications of designing generative models that are similar or dissimilar from the generative process (Hohwy 2013; Clark 2015; Pezzulo, Donnarumma et al. 2017; Nave et al. 2020, Tschantz et al. 2020). As discussed above, the *good regulator theorem* (Conant and Ashby 1970) says that an effective adaptive creature must *have* or *be* a good model of the system it regulates. However, this can be achieved in various ways. First, as discussed so far, the creature's generative model can mimic (at least to a great extent) the generative process. Models developed in this way may be called *explicit* or *environmental* models, given the resemblance between their internal states and the environment's external states. Second, the creature's generative model can be much more parsimonious than (and even significantly different from) the generative process, to the extent that it correctly manages those aspects of the environment that are useful to act adaptively in it and achieve the creature's goals. Models developed in this way may be called *sensorimotor* or *action oriented*, as they mostly encode action-observation (or sensorimotor) contingencies and their primary role is supporting goal-directed actions as opposed to providing an accurate description of the environment.

The difference between explicit and action-oriented models can be appreciated if we consider different ways one can model (for example) a rodent trying to escape from a maze in which some corridors are dead ends. An explicit generative model may resemble a cognitive map of the maze and provide a detailed characterization of external entities, such as specific locations, corridors, and dead ends. This model may permit the rodent to escape from the maze using map-based navigation. An action-oriented model may instead encode contingencies between whisker movements and touch sensations. This latter model would afford the selection of contextually appropriate strategies, such as moving forward (if no touch sensation is experienced or expected) or changing direction (in the opposite case)—eventually permitting the rodent to escape from the maze without explicitly representing locations, corridors, or dead ends. These two kinds of model prompt different philosophical interpretations of Active Inference, considering generative models as ways to either reconstruct the external environment (explicit) or afford accurate action control (action oriented).

Finally, as discussed in the field of *morphological computation* (Pfeifer and Bongard 2006), some aspects of a creature's or a robot's control can be outsourced to the body and hence do not need to be encoded in its generative model. One example is the *passive dynamic walker*: a physical object resembling a human body, composed of two "legs" and two "arms," which is able to walk an incline with no sensors, motors, or controllers (Collins et al.

2016). This example implies that at least some aspects of locomotion (or other abilities) can be achieved with body mechanics that are carefully tuned to exploit environmental contingencies (e.g., an appropriate body weight or size to walk without slipping); therefore, these contingencies do not need to be encoded in the creature's generative model. This suggests an alternative way to design Active Inference agents (and their bodies) that *are*—as opposed to *have*—good models of their environment. Yet all the ways to design Active Inference models are not mutually alternative but can be appropriately combined, depending on the problem of interest.

6.7   Simulating, Visualizing, Analyzing, and Fitting Data Using Active Inference

In most practical applications, once the generative model and generative process have been defined, one only needs to use the standard procedure of Active Inference—the descent of the active and internal states on the free energy functional associated with the model—to obtain numerical results. Arguably, modelers' goals are to simulate, visualize, analyze, and fit data (e.g., conduct model-based data analysis). Standard routines for Active Inference that provide support for all these functions are freely available (https://www.fil.ion.ucl.ac.uk/spm/); an annotated example of using these routines is provided in appendix C.

Although in most cases Active Inference procedures function off-the-shelf, in some practical applications one may consider specific fine-tunings or changes. For example, specifying the temporal depth of planning defines how many future states are considered during expected free energy computations. Setting up a limited temporal depth, along with other approximations to exhaustive search such as sampling (Fountas et al. 2020), may be useful in practical applications of Active Inference in large state-spaces.

Another example of adapting the standard functioning of Active Inference is the selective removal of parts of the expected free energy equation. This ablation may be useful to compare standard Active Inference (that uses expected free energy) with reduced versions, in which some parts of the expected free energy are suppressed to render them formally analogous to (for example) KL control or utility maximization systems (Friston, Rigoli et al. 2015). Furthermore, one can also augment Active Inference models with additional mechanisms, such as habitual learning (Friston, FitzGerald et al. 2016) or learning rate modulation (Sales et al. 2019), with the caveat that maintaining the normative character of Active Inference would require casting these additional mechanisms in terms of free energy minimization.

Finally, other fine-tunings or changes to Active Inference may be useful to characterize disorders of inference and psychopathological conditions—for example, to explore the behavioral and neuronal consequences of endowing a creature's generative model with excessively strong (or weak) priors via excessively high (or low) levels of neuromodulators. We will provide some examples of Active Inference models that are relevant for psychopathology in chapter 9.

6.8   Summary

In this chapter, we have outlined the most important design choices that must be made in setting up an Active Inference model. We provided a recipe in four steps and some guidelines to address the usual challenges that model designers face. Of course, it is not necessary to follow the recipe in a rigid manner. Some steps can be inverted (e.g., design the generative process before the generative model) or combined. But in general, these steps are all required. This sets up the remainder of this book, which puts these ideas into practice through a series of illustrative examples designed to showcase the theoretical principles presented in the first half of the book. In everything that follows, the only differences among the examples rest on the design choices we have highlighted here. Part 2 illustrates systems with different boundaries, with discrete or continuous dynamics at different timescales, for which the choice of prior beliefs is fundamental in reproducing behavior across many different domains—but all implementing the same Active Inference.

7 Active Inference in Discrete Time

What I cannot create, I do not understand.

—Richard Feynman

7.1 Introduction

So far, we have discussed the principles of Active Inference at a relatively abstract level. This chapter deals with specific examples—and how they may be specified in a practical setting. We focus on models of categorical variables in discrete time. Through a series of examples, building in complexity, we illustrate models of perceptual processing, decision-making, information seeking, learning, and hierarchical inference. These examples are chosen to highlight as simply as possible emergent properties—including measurable physiology and behavior—of Active Inference schemes.

7.2 Perceptual Processing

We begin by considering perceptual processing and the inversion of the sort of discrete-time models introduced in chapter 4. Later in this chapter, we build to a full partially observable Markov decision process (POMDP). However, we start with a special case of a POMDP in which we can ignore choices and behavior: a hidden Markov model (HMM), which may be used for perceptual inference of a sequential and categorical sort (see figure 7.1). To motivate this, we will appeal to a simple example. Imagine listening to a performance of a short piece of music. The sequence of notes that are written in the score may be thought of as hidden (unobserved) states, while the sequence of notes we actually hear are the (observable) outcomes. If the performer is a professional musician, the correspondence between the

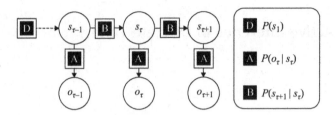

Figure 7.1

This hidden Markov model uses the same notation introduced in chapter 4 to express a sequence of states (*s*) that evolve through time. At each time, they give rise to an observable outcome (*o*). The state at one time depends only on the state at the previous time (with this dependency expressed in **B**). The first state in the sequence has prior probability **D**. The generation of outcomes from states depends on the likelihood distribution (**A**). This specification of an HMM is generic, with specific generative models depending on specific choices for **A**, **B**, and **D**.

hidden states and the outcomes may be very close. However, if an amateur, there may be an additional degree of stochasticity in the (likelihood) mapping from the note that should be played to that which is heard. In this scenario, it may still be possible to infer which note should have been heard, given prior beliefs about the probability that each note is preceded or succeeded by another.

The example of listening to the amateur musician may be formalized in the following way. First, we decide on how reliably our musician actually plays the note (outcome) she intends to (hidden state). We can express this through the **A**-matrix, whose elements indicate the probability of an outcome (rows) given a state (columns). In our toy example, we set this as follows:

$$A = \frac{1}{10} \begin{bmatrix} 7 & 1 & 1 & 1 \\ 1 & 7 & 1 & 1 \\ 1 & 1 & 7 & 1 \\ 1 & 1 & 1 & 7 \end{bmatrix} \tag{7.1}$$

This says that 70 percent of the time, our musician hits her intended note. We then specify the transition probabilities in the **B**-matrix, which account for the probability of the next state (rows) given the current state (columns):

$$B = \frac{1}{100} \begin{bmatrix} 1 & 1 & 1 & 97 \\ 97 & 1 & 1 & 1 \\ 1 & 97 & 1 & 1 \\ 1 & 1 & 97 & 1 \end{bmatrix} \tag{7.2}$$

This says that there is a 97 percent probability of the first note being followed by the second, the second by the third, and so on. If we know that the sequence always begins with the first note, we set the prior probability:

$$D = \begin{bmatrix} 1 \\ 0 \\ 0 \\ 0 \end{bmatrix} \tag{7.3}$$

Together, equations 7.1–7.3 completely specify the HMM generative model shown in figure 7.1. In other words, they provide a description of our beliefs about how the music we hear is generated by our amateur musician. Using equation 4.12 and substituting in our generative model, we can simulate the dynamics of the Bayesian belief updating induced by a sequence of outcomes. This is shown in figure 7.2. Note the increase in confidence shown in the upper-left plot as more data are accumulated over time, except for the third time step, where an unexpected outcome has occurred. This outcome could be explained in two ways. First, it may be that the intended note really was an unusual note under our prior beliefs in equation 7.2. This is made less likely by the rarity of such transitions under the **B**-matrix of this model. The alternative, more plausible explanation is that the musician played the wrong note by mistake. As shown in the third column of the upper-right plot, this is the explanation that our simulated listener settles on. However, a nonzero probability is assigned to the possibility that it was the right note after all. The capacity to report this sort of uncertainty is a key feature of the Bayesian perspective afforded by Active Inference.

The model shown here may be made more sophisticated in many ways, but perhaps the simplest relies on the factorization of the state-space (Mirza et al. 2016). An example might be the pitch and dynamics of the note (with a similar distinction in the outcomes). In a visual inference task, the factorization may be into what and where, which has a great deal of currency in neurobiology (Ungerleider and Haxby 1994). In subsequent sections, we will appeal to this sort of factorization to separate those states that can be influenced by the creature in question from those that cannot. For further reading on this sort of model (without actions in play) and the kinds of neuronal message passing scheme that might be used to invert it through minimizing free energy, see Parr, Markovic et al. (2019).

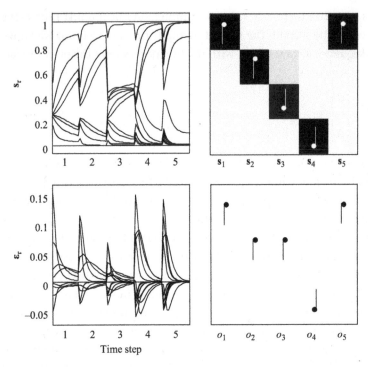

Figure 7.2
These simulated perceptual inference plots illustrate the belief-updating process in an example trial based on the generative model outlined in the main text. *Upper-left:* Beliefs (posterior probabilities) about each note in the sequence at each time step. *Upper-right:* As the numerical values of these beliefs are difficult to track the beliefs at the end of the sequence, having heard each note (i.e., retrospective beliefs) are shown. Each column shows (retrospective) beliefs about the hidden states at a given time step. Each row represents an alternative hypothesis for that hidden state. The darker the shading, the more probable that note is considered to have been (with black indicating a probability of one and white a probability of zero). *Lower-left:* (Negative) free energy gradients (i.e., prediction errors) over time. The rate of change of the beliefs in the upper-left plot is determined by the value of these errors at each time step. *Lower-right:* Sequence of musical notes presented to our synthetic agent (i.e., the observations he receives during time steps 1 to 5). Note that while at the third time step (o_3) the listener heard the second note (third column of the lower-right plot), he infers the third note with higher probability (third column of the upper-right plot).

7.3   Decision-Making and Planning as Inference

The HMM used above illustrates a very simple form of categorical inference based on a sequence of outcomes. However, the sort of (sessile) creature that this describes is rather uninteresting. Autonomous creatures are clearly more than passive recipients of sensory data. Instead, they actively change their environment and engage in a bidirectional exchange with their sensorium. This speaks to the importance of converting an HMM into a POMDP, whereby we must infer not only how our environment is changing but also how our chosen course of action changes it and which course of action to choose.

Figure 7.3 shows a POMDP generative model. This is the same as that introduced in chapter 4, where the details of inference in this sort of model are unpacked. Note the similarity of this structure to the HMM in figure 7.1

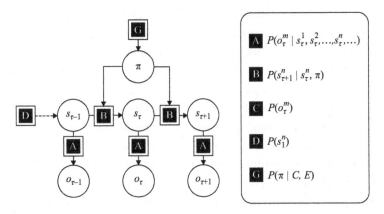

Figure 7.3
POMDP from figure 4.3, unpacking the probability distributions in terms of hidden state factors and outcome modalities. (Figure 7.1 is a special case of this structure.) Three points of note: First, the factorization of the hidden states now means that the distribution encoded by **A** has (potentially) many state factors in its conditioning set and can no longer be encoded by a matrix. Instead, this becomes a tensor object, in which each index corresponds to a state factor. Second, the separation of the outcomes into different modalities means there will be a separate **A** tensor for each modality. Third, while **C** and **E** appear in the panel on the right, they do not appear in the factor graph on the left because they only get into the generative model via prior beliefs about policies. For an alternative perspective on this, see Parr and Friston (2018d) and van de Laar and de Vries (2019).

and the addition of an extra variable (π), on which the transition probabilities (**B**) are conditioned. This means we can entertain alternative hypotheses about the dynamics of states. These hypotheses may be interpreted as plans that a creature may select between. This perspective equates policy evaluation with model comparison and says that a policy is simply an explanatory variable for an observed sequence of (self-generated) sensations.

The model in figure 7.3 differs subtly from that introduced in chapter 4: it allows factorization of states (superscript n) and of outcomes (superscript m). The utility of this is obvious when we consider the factorization of the visual world into where an object is and what it is. Clearly, it would be extremely inefficient (and incur a high complexity cost) to represent every possible combination of location and identity, when identity is (normally) invariant to location and vice versa. A similar argument may be used for factorization of time from identity and location (Friston and Buzsaki 2016). The benefit of introducing this factorization at this stage is that we can separate those states of the world over which a creature has control from those that it does not. While the transition probabilities governing the former will be different under each policy, the latter will be invariant to this.

With these preliminaries in place, we now outline a simple example of a task (Friston, FitzGerald et al. 2017) that requires planning and illustrates some of the key aspects of active inference using POMDPs. This involves a rat in a T-maze containing an aversive stimulus in one arm, an attractive stimulus in another, and a cue that indicates the location of the two stimuli in the final arm. This setup means that the rat can behave in (broadly) two ways. It could choose to go straight to one of the two arms that might contain the attractive stimulus, risking the aversive stimulus. Alternatively, it could choose to seek out the informative cue and then go to the arm most likely to contain the attractive stimulus.

This choice speaks to the classical exploration-exploitation dilemma in psychology: a dilemma that is resolved under Active Inference. The resolution stems from the minimization of expected free energy mandated by prior beliefs about policies. To review this briefly (see chapter 4 for details), the most probable policies (for a creature who minimizes its variational free energy) are those that lead to the lowest expected free energy. The expected free energy has the following form:

$$G(\pi) = \underbrace{\mathbb{E}_{Q(\tilde{s}|\pi)}[H[P(\tilde{o}|\tilde{s})]] - H[Q(\tilde{o}|\pi)]}_{\text{Negative epistemic value } (-\mathcal{I}(\pi))} - \underbrace{\mathbb{E}_{Q(\tilde{o}|\pi)}[\ln P(\tilde{o}|C)]}_{\text{Pragmatic value}} \qquad (7.4)$$

This decomposition of the expected free energy into epistemic and pragmatic value highlights the (epistemic) drive toward information gathering and the (pragmatic) drive toward realizing prior beliefs (**C** in figure 7.3). We will attempt to provide a deeper intuition for the *epistemic value* in the next section, but it can be thought of simply as the amount of information we stand to gain under a specific policy. The form of the *pragmatic value* effectively treats the probability of outcomes, averaged over all policies, as if it were a prior. In doing so, those policies with consequences consistent with this prior become more probable, as they are associated with lower expected free energy. To put this in more intuitive terms, if we consider a certain sort of observation to be very probable, we will act to fulfill our belief that we will encounter these. Therefore, the log probability of outcomes may be thought of as equivalent to a utility function in other formalisms, such as optimal control theory and reinforcement learning. The fact that utility and the value of information emerge as two components of the expected free energy means that we do not need to worry about balancing exploration and exploitation. Both are in service of optimizing the same function.

To see how this unfolds in the T-maze example, we need to formalize the generative model in the same way as with the HMM above. Figures 7.4–7.6 illustrate the likelihood and transition probabilities that comprise the generative model for the T-maze. We will go through this in some detail, as this minimal example provides the building blocks from which readers can construct their own generative models. The first thing to do is to decide on the number of outcome modalities that represent the (sensory) data our model is supposed to explain. This tells us the number of **A**-matrices we must specify. Here, we have two modalities that represent exteroceptive data pertaining to *where* the rat is in the maze ($\mathbf{A}^1$) and a *what* modality that may be the interoceptive data the rat experiences when it has found the attractive (edible) stimulus ($\mathbf{A}^2$). The levels in these modalities (i.e., the alternative observations that could be made in each) determine the rows of each **A**-matrix. The next decision is the number of hidden state factors that may be used to explain these data; this is the number of **B**-matrices we require. We consider two factors here: the position of the rat in the maze, and the context (attractive stimulus on left or right). These have four and two levels, respectively. We now must specify, for each combination of hidden states, the probability of each outcome. Context 1 is shown in figure 7.4; context 2 is shown in figure 7.5. For the first modality, our $\mathbf{A}^1$ associates each location with an outcome with

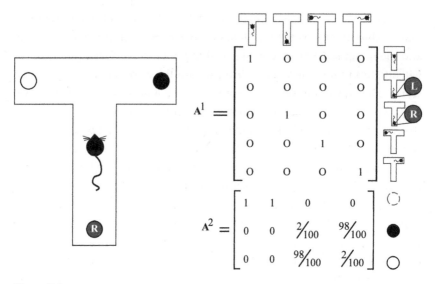

Figure 7.4

Likelihood in context 1. *Left:* T-maze configuration of cues and stimuli: the attractive stimulus is on the right and the aversive stimulus is on the left. *Right:* Likelihood or observation model specifies the probabilistic mapping from location to exteroceptive cues (A^1) and to interoceptive cues (A^2). Each element of these matrices is the probability of the outcome illustrated at the end of the row, conditioned on the context being one, and on being in the location indicated by the row. The exteroceptive outcomes are visual or proprioceptive input associated with each location, whereby the cue location can give rise to a rightward or a leftward cue. The interoceptive outcomes are absent (circle with dashed outline), attractive (filled circle), or aversive (unfilled circle).

probability one. The cue location may be associated with a left or a right cue, depending on the context. The interoceptive modality (A^2) associates a neutral outcome with the start and cue locations and a 98 percent chance of finding the attractive outcome when the context matches the arm of the maze the rat has entered. Technically, these **A**-matrices are tensor quantities, because their elements are specified by three numbers (outcome, location, and context), while a matrix is only specified by two (row and column).

We then need to specify transition probabilities. The **B**-matrices specify the probability of transitioning from a state (column) to another state (row), depending on the choice of policy (π). These specify the transitions pertaining to the position of the rat in the maze (B^1) and transitions in the context

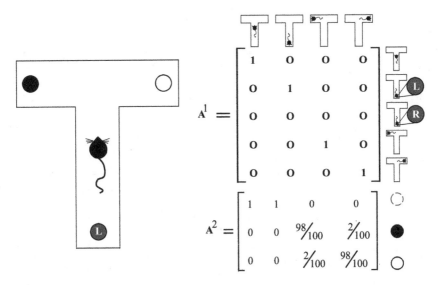

Figure 7.5
Likelihood in context 2. Nearly identical to figure 7.4—in this context, the aversive and attractive stimuli have been swapped. This is reflected in the probability of the exteroceptive outcomes in the cue location and the probabilities of the interoceptive outcomes in the right and left arms of the maze.

(B^2). Figure 7.6 shows the controllable B^1-transitions. Each matrix shows the probabilities under a different action choice (subscripted). These allow a move from any location to any other location, except for from the two arms of the maze, which are absorbing states. This means that once there, the rat must stay there, regardless of the actions it chooses. In contrast, the rat has no control over the context (i.e., whether it is in context 1, shown in figure 7.4, or context 2, shown in figure 7.5). Context stays constant over time and can be represented as an identity matrix:

$$B_\pi^2 = \begin{bmatrix} 1 & 0 \\ 0 & 1 \end{bmatrix} \tag{7.5}$$

Here each column (and row) refers to a state indexing either figure 7.4 or figure 7.5. This means that whichever context we start in stays constant (transitions to itself) over time. This is true regardless of the policy selected. The C^1-vector shows prior preferences for each of the outcomes in this modality, with uniform preferences except for a slight aversion (−1) to

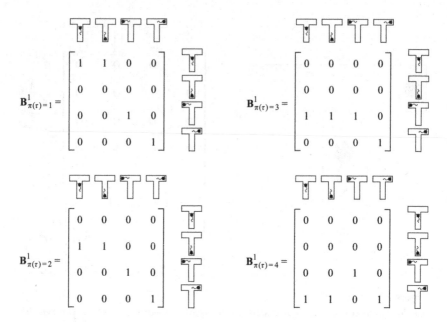

Figure 7.6
Controllable transition probabilities for moving between the different locations. Each of the four matrices corresponds to an alternative action the rat can choose. These allow for a move from any state (except the right and left arm) to any other state. The right and left arms are absorbing states, in which the rat must stay once entered.

the start location. The $\mathbf{C}^2$-vector specifies preferences (+6) for the attractive stimulus and aversion (–6) to the aversive stimulus. The absence of either is considered neutral (0).

$$\mathbf{C}^1 = \sigma\big([-1,0,0,0,0]^T\big)$$
$$\mathbf{C}^2 = \sigma\big([0,6,-6]^T\big)$$

(7.6)

The order of elements in these vectors corresponds to the order of rows in the corresponding **A**-matrices. The softmax function (σ) allows us to specify preferences in terms of positive and negative values (corresponding to unnormalized log probabilities), which are then converted to probabilities. This preserves the difference in log probabilities (or the relative probability) while ensuring normalization. Practically, this formulation means the attractive stimulus is considered e^6 (≈ 400) times more probable than the neutral stimulus under the rat's generative model. This is a very strong preference that means the rat believes its actions are much more likely to lead to the attractive outcome. This constraint on inference about action is crucial for the behavior that follows. Finally, the **D**-vectors specify the prior probabilities for the initial states:

$$\mathbf{D}^1 = [1, 0, 0, 0]^T$$
$$\mathbf{D}^2 = \tfrac{1}{2}[1, 1]^T$$

(7.7)

The order of elements in these vectors matches those of the **B**-matrices. The **D**^{1}-vector indicates a confident belief in starting at the center of the maze. The **D**^{2}-vector indicates that the two contexts (figure 7.4 or 7.5) are considered equally probable at the start.

Figure 7.7 shows what happens when we invert the generative model of figures 7.4–7.6. The upper row illustrates what we would see if observing the rat's behavior. It starts in the center and then goes to the informative cue. This is due to the high *epistemic value* associated with this location (i.e., the observations made at this location have the potential to resolve uncertainty about the context). On seeing the cue that indicates a left context (context 1), the rat chooses the left arm of the maze and finds the rewarding stimulus. This move is driven by the high *pragmatic value* attributed to this location. The lower plots illustrate the belief updating that occurs during this simple trial. As in figure 7.2, this is shown in the form we might expect to observe in an idealized rat if we were measuring neuronal activity (i.e., firing rates and local field potentials [LFPs]). Note the rapid change in beliefs at the second time step, when the rat reaches the informative cue location, and associated LFP (dashed line).

7.4 Information Seeking

The simulation in section 7.2 illustrates a simple example of an exploration-exploitation trade-off, which is solved by foraging for information until uncertainty is resolved, then exploiting what has been inferred to fulfill prior preferences. In this section, we unpack the concept of epistemic value in greater detail. As we saw in equation 7.4, this comprises two terms:

$$\underbrace{\mathcal{I}(\pi)}_{\text{Epistemic value}} = \underbrace{H[Q(\tilde{o}|\pi)]}_{\text{Post. pred. entropy}} - \underbrace{\mathbb{E}_{Q(\tilde{s}|\pi)}[H[P(\tilde{o}|\tilde{s})]]}_{\text{Expected ambiguity}}$$

$$= \underbrace{D_{KL}[P(\tilde{o}|\tilde{s})Q(\tilde{s}|\pi) \, \| \, Q(\tilde{o}|\pi)Q(\tilde{s}|\pi)]}_{\text{Mutual information}}$$

(7.8)

$$= \underbrace{\mathbb{E}_{Q(\tilde{o}|\pi)}[D_{KL}[Q(\tilde{s}|\pi,\tilde{o}) \, \| \, Q(\tilde{s}|\pi)]]}_{\text{Information gain, salience, Bayesian surprise}}; \quad Q(\tilde{s}|\pi,\tilde{o}) \triangleq \frac{P(\tilde{o}|\tilde{s})Q(\tilde{s}|\pi)}{Q(\tilde{o}|\pi)}$$

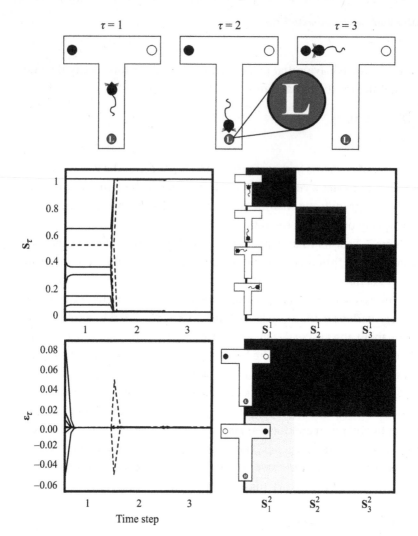

Figure 7.7

Simulated epistemic and pragmatic behavior of a rat foraging in a T-maze. The rat starts in the central location but then chooses to sample the informative cue in the lower arm of the maze. This location is associated with the greatest *epistemic value*, as observing the cue in this location reveals the context (reward right or left) that the rat finds itself in. On observing the cue, the rat undergoes rapid belief updating (s), inducing an LFP (ε). With no more uncertainty to resolve, the rat selects the *pragmatically valuable* option and goes to the left arm of the maze. The two plots on the right show the beliefs held by the rat at the end of the trial about all previous times (i.e., these are retrospective beliefs and not the beliefs of the rat at the moment of the decision). It believes (correctly) that it started in the central location, went to the cue arm, and then went to the left arm. For the context hidden state factor, the rat believes that the context was the left context throughout.

These are the *posterior predictive entropy* and the *expected ambiguity*, respectively. Below these, we highlight the correspondence between these and other rearrangements. To unpack these in an intuitive way, we will frame this in terms of a visual paradigm, where alternative saccades (π) lead to different transitions between fixation locations (s). In addition to fixation locations, the hidden states include the identity of a stimulus at each location. A combination of stimulus and fixation generate visual and proprioceptive consequences (o). With this in mind, we can interpret the posterior predictive entropy as the dispersion (or uncertainty) associated with "what I would see if I performed this eye movement." From the perspective of a scientist, this quantifies how uncertain we might be about the data we would obtain on performing a given experiment. Under this perspective, it makes sense that we should select those saccades (or experiments) that are associated with the greatest posterior predictive entropy, as these offer the greatest potential for uncertainty resolution. We would gain nothing by performing an experiment if we already knew what the results would be with a high degree of confidence.

However, the predictive entropy only tells us the total amount of uncertainty. It does not tell us how much uncertainty is actually resolvable. We will always be uncertain about the next number in a sequence of randomly generated numbers, but we will never resolve our uncertainty about the process generating them by fixating on these. This is where the expected ambiguity comes in. This quantifies the degree to which observations and states are independent of one another. If states always generate the same observation, this quantity will be zero. It will be maximal if, as in the random number generator, there is no association between states and outcomes. In the visual domain, this implies that the best saccade will be that toward a well-lit stimulus, where there is little ambiguity about "what I would see if I looked at this stimulus." Taken together, this says that the best saccades (i.e., perceptual experiments) are those for which there is the greatest uncertainty to resolve (posterior predictive entropy) but only if that uncertainty can be resolved (negative ambiguity). Interestingly, this has exactly the same form as expressions developed in statistics to score experimental design in terms of information gain (Lindley 1956).

Figure 7.9 illustrates what happens in a saccadic paradigm (Parr and Friston 2017b) when we simulate manipulations to the ambiguity and posterior predictive entropy. This shows four stimuli (squares), each of which may change color from moment to moment. Superimposed on these is a simulated eye-tracking trace, as if we were measuring where an experimental participant was looking. Crucially, we specify prior beliefs about outcomes to be uniform (i.e., pragmatic value to be absent), precluding any preference-based choices. This means each saccade is selected to maximize epistemic value. When the generative model treats all four stimuli as equivalent (left image), all are sampled with approximately the same frequency. However, we can modulate the uncertainty associated with each stimulus (see box 7.1). If we set one stimulus to have a greater ambiguity (by increasing the value of off-diagonal elements of the corresponding **A**-matrix), this square is ignored (middle image). This is an example of the famous "streetlight" effect (Demirdjian et al. 2005), which takes its name from the metaphor of

Box 7.1
Uncertainty and precision

The example in figure 7.7 appeals to the concept of precision—an important idea in this book. Precision is the inverse of variance and scores our confidence in a given probability distribution. This is closely related to the negative entropy (*negentropy*) of a distribution:

$$-H[P(s)] = \mathbb{E}_{P(s)}[\ln P(s)]$$

A simple way to parameterize a distribution such that it can be made more or less precise is to use a Gibbs form with an *inverse temperature* parameter (ω). This has the following form:

$$P(s|\omega) = Cat(\sigma(\omega \ln \mathbf{D}))$$

Note that the precision multiplies the log prior, so it can be interpreted as a gain-control device (amplifying as opposed to adding to neural signals). The plots in figure 7.8 show how the probability distribution (each column representing the probability of an alternative state) changes for a given **D** when we vary ω. Note the increasing confidence with increasing precision.

This sort of parameterization may be applied to any of the distributions used in a POMDP. In addition, we can define priors over the precision and infer

$$\omega = 0.1 \qquad \omega = 1 \qquad \omega = 10 \qquad \omega = 100$$

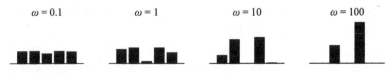

Figure 7.8

Box 7.1 (continued)

this just as we infer other latent variables (i.e., through free energy minimization). Assuming the prior has a Gamma distribution (precluding negative values of the precision), we get the following updates (see appendix B for details):

$$P(\omega) = \Gamma(1, \beta_\omega)$$
$$Q(\omega) = \Gamma(1, \boldsymbol{\beta}_\omega)$$
$$\Rightarrow \dot{\boldsymbol{\beta}}_\omega = (\mathbf{D}^{\beta_\omega^{-1}} - \mathbf{s}) \cdot \ln \mathbf{D} + \beta_\omega - \boldsymbol{\beta}_\omega$$

There is an increasing recognition that the biological substrate of these precision parameters may be the neuromodulatory systems that set the gain of neural responses. Chapter 5 discusses the evidence relating these parameters to specific neurochemicals.

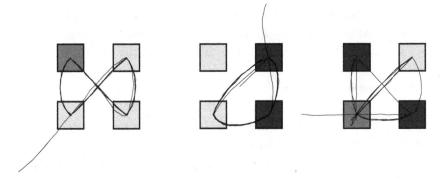

Figure 7.9
Simulated epistemic visual search paradigm (Parr and Friston 2017b) with the synthetic eye-tracking trace superimposed on the four stimulus locations. Each stimulus (shaded square) is associated with a transition matrix that may be more or less predictable and a likelihood matrix that may be more or less ambiguous. *Left:* When transitions and likelihoods are equally predictable for all four locations, all locations are sampled with about the same frequency. *Middle:* The viewer shows aversion to the upper-left square when it is specified with a less precise (more ambiguous) likelihood mapping. *Right:* The lower-left square is epistemically attractive when the transition probabilities are specified as more uncertain.

people who have lost their keys late at night. The first place they might look is under the streetlight—not because the keys are most likely to be there, but because it is the best place to find high-quality, unambiguous, uncertainty-resolving information. The simulation shows how the ambiguous (e.g., poorly lit) square is ignored, reproducing an in silico streetlight effect.

In contrast, the right image in figure 7.8 shows what happens when we make the transitions less predictable for the lower-left square. We accumulate uncertainty about this location very quickly, ensuring a high posterior predictive entropy with no change to the ambiguity. As we can see, this leads to more frequent fixation on this location, as there is always new uncertainty to resolve here. Intuitively, if I know something has very predictable dynamics, I do not have to look at it very often to be confident about its state. In contrast, if something may have changed in the time that I have been looking at something else, it is worth looking back at to check. These simulations are designed to offer an intuition for the two parts of the epistemic value, to see how minimization of expected free energy ensures we actively select our sensory data to find out about the world.

7.5   Learning and Novelty

Sections 7.2–7.4 set out everything that is required for the majority of practical applications of Active Inference. However, we have assumed that the generative model is already known and does not change as an effect of learning. In some practical applications, we may want to consider how one or more parts of the generative model (e.g., the **A**- or **B**-matrix) are learned during an experiment or, more broadly, how we optimize the structure of the generative model itself, given some data (Friston, FitzGerald et al. 2016). In doing so, Active Inference extends to active learning, and the *salience* (equation 7.8) describing information gain about states is complemented by *novelty*, which deals with resolution of uncertainty about (for example) the elements of the **A** matrix shown in equation 7.1, the **B** matrix shown in equation 7.2, or any other parameters of the generative model. These beliefs can now vary with time rather than being fixed, as assumed so far (Schwartenbeck et al. 2019). To get to this, we first have to extend the generative model as in Figure 7.10 to include beliefs about these model parameters.

Conceptually, including beliefs about parameters in the generative model permits treating learning as another form of Bayesian inference—namely,

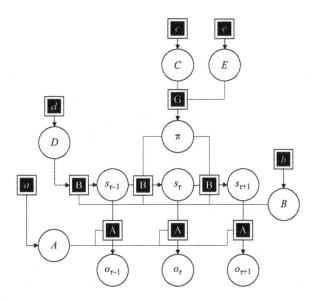

Figure 7.10
This generative model for learning uses the same POMDP structure as in figure 7.3, but the priors for each of the hidden states now depend on variables (in circles), which themselves now come equipped with prior beliefs. These have the form of Dirichlet distributions, which are conjugate (see box 7.2) to the categorical distributions considered thus far. The model shows how the likelihood of outcomes given states now also depends on a variable A (which is the same for all time-points), the transition probabilities are now conditioned on a variable B, the preferences depend on C, the initial states depend on D, and the fixed form policy prior depends on E. By making prior beliefs about the parameters of the generative model explicit, this figure emphasizes that both inference and learning are free energy minimizing processes, but they are distinct. In short, inference describes the optimization of beliefs about the state of the world as it is (s), including beliefs about the way in which we are acting (π). In contrast, learning describes optimization of beliefs about the relationships between these variables ($A, B, C, D,$ or E). The latter vary much more slowly than the former and may only be learned when the states have been inferred. We will return to this separation of timescales below when we consider hierarchical generative models.

as the passage from prior to posterior beliefs about model parameters. This highlights the fundamental similarity of perception and learning: in the same way that perception can be described as the inversion of a generative model to infer hidden states from observations, learning can be described as the inversion of a generative model to include beliefs about parameters (although normally this inversion may operate on a slower timescale).

Box 7.2

Conjugate priors

> When setting up a generative model of the form in figure 7.10, it is important to carefully select the appropriate distribution for prior beliefs. Typically, these will be the *conjugate* prior distribution associated with the likelihood. A conjugate prior belief means that, when used to perform Bayesian inference, the posterior belief will be the same type of distribution. For example, using Bayes' rule:
>
> $$P(D|s) \propto P(D)\, P(s|D)$$
>
> If $P(s|D)$ is a *categorical* distribution, when we choose a *Dirichlet* distribution (conjugate to *categorical*) for $P(D)$, we can guarantee that $P(D|s)$ is also a *Dirichlet* distribution. Put formally:
>
> $$\left. \begin{array}{l} P(D) = Dir(d) \\ P(s|D) = Cat(D) \end{array} \right\} \Rightarrow P(D|s) = Dir(\mathbf{d})$$

The simplest way to choose the right kind of prior is to look up the conjugate prior for whatever form the likelihood distribution takes. For the categorical distributions used here, a Dirichlet distribution is the appropriate choice for beliefs about parameters (see box 7.2). Having included these additional prior beliefs, we can now optimize posterior beliefs about the structure of the generative model. This means incorporating these into the free energy (as we did for states in chapter 4) and finding the free energy minima.

$$\theta = (A, B, C, D, E)$$
$$F = \mathbb{E}_{Q(\pi,\theta)}[F(\pi,\theta)] + D_{KL}[Q(\theta) \| P(\theta)] + D_{KL}[Q(\pi) \| P(\pi)] \tag{7.9}$$

Dirichlet distributions are parameterized by counts (or pseudo-counts) that index the number of times a given categorical variable has been seen (or, in the case of the priors, as if it had been seen that number of times). For the derivation of the update rules for these parameters, see appendix B. For now, we summarize the update rule and key properties of a Dirichlet distribution, focusing on the a and $\mathbf{a}$ concentration parameters associated with the prior and posterior over A.

$$\mathbf{a} = a + \sum_{\tau} s_{\tau} \otimes o_{\tau}$$
$$\mathbb{E}_Q[A_{ij}] = \mathbf{A}_{ij} \approx \frac{\mathbf{a}_{ij}}{\mathbf{a}_{0j}}$$
$$\mathbb{E}_Q[\ln A_{ij}] = \ln \mathbf{A}_{ij} = \psi(\mathbf{a}_{ij}) - \psi(\mathbf{a}_{0j}) \tag{7.10}$$
$$\mathbf{a}_{0j} \triangleq \sum_{k} \mathbf{a}_{kj}$$

The first line here expresses the update from prior to posterior concentration parameters following a series of observations, with beliefs about the states that caused them. The cross in the circle indicates a Kronecker tensor product (or outer product in the case of two vectors), here giving rise to a matrix in which each element is the product of a pair of elements in s_τ and o_τ. This update rule may be interpreted simply as a form of activity-dependent plasticity. When an outcome is observed in combination with a posterior belief that a particular state caused it, the element of the matrix representing the relationship between the two is incremented. The second line of the equation highlights the interpretation of the Dirichlet concentration parameters in terms of counts. For a given state (column), each element of **a** is the number of times the corresponding outcome has been seen. Dividing by the sum of the elements in the column (total number of observations or pseudo-observations) gives the probability of each outcome given that state. To understand why this (pseudo) counting method makes intuitive sense, consider the amateur musician example from the beginning of this chapter. If one counts how many times the musician hits the first note when she intends to do so (first row and column), how many times she hits the second note when she intends to do so (second row and column), and so on, and divides these by the total number of times she intends each note, one will eventually converge to the correct numerical values of the **A**-matrix shown in equation 7.1—namely, that the musician hits all her intended notes 70 percent of the time. The counting method has another important consequence that we will return to: The number of counts or pseudo-counts preceding an observation tells us how likely we are to update our beliefs on making the observation. Imagine flipping a coin five times and getting five heads in a row. This might lead us to update our beliefs to favor the hypothesis that this is an unfair coin. However, if this had been preceded by 100 flips with 50 heads and 50 tails, the final five heads would do little to influence our beliefs about whether this is a fair coin. The third line of equation 7.10 shows a useful identity associated with Dirichlet distributions: the expected log of the random variable is given by the difference in two digamma functions (derivative of a gamma function).

The inferential approach to learning highlights an important difference between Active Inference and most other approaches to computational neuroscience and machine learning, which incorporate various learning rules (e.g., Hebbian rules or error backpropagation) that are considered biologically realistic or computationally efficient. In Active Inference, the update rules that govern learning are derived from statistical consider-ations, yet they turn out to be remarkably similar to biologically motivated rules for activity-dependent plasticity (see the above considerations on the first line of equation 7.10). This exemplifies one of the appeals of norma-tive approaches, which start from first principles to explain what we know about brains and behavior—and things that we did not know.

A further difference between Active Inference and most machine learn-ing approaches is that learning is naturally described as an *active* process, in which creatures autonomously select the most appropriate data to improve their generative models. This becomes evident if one considers that when including beliefs about parameters in the model, the expected free energy acquires an additional term:

$$
G(\pi) = \underbrace{D_{KL}[Q(\tilde{o}|\pi) \,\|\, P(\tilde{o}|C)]}_{\text{Risk}} + \underbrace{\mathbb{E}_{Q(\tilde{s}|\pi)}[H[P(\tilde{o}|\tilde{s})]]}_{\text{Ambiguity}}
$$

$$
+ \underbrace{\mathbb{E}_{\tilde{Q}(\tilde{o},\tilde{s},\theta|\pi)}[\ln Q(\theta) - \ln P(\theta|\tilde{o},\tilde{s})]}_{\text{Parameter information gain}} \tag{7.11}
$$

$$
= -\underbrace{\mathbb{E}_{Q(\tilde{o}|\pi)}[D_{KL}[Q(\tilde{s}|\pi,\tilde{o}) \,\|\, Q(\tilde{s}|\pi)]]}_{\text{Salience}}
$$

$$
- \underbrace{\mathbb{E}_{\tilde{Q}(\tilde{o},\tilde{s}|\pi)}[D_{KL}[Q(\theta|\tilde{o},\tilde{s}) \,\|\, Q(\theta)]]}_{\text{Novelty}} - \underbrace{\mathbb{E}_{Q(\tilde{o}|\pi)}[\ln P(\tilde{o}|C)]}_{\text{Pragmatic value}}
$$

The *salience* and *pragmatic value* terms were already in place in equation 7.4, but the *novelty* term is new. The final equality here shows an arrangement that highlights the relationship between salience and novelty. In short, salience is to inference what novelty is to learning. Both are expressions of the change in beliefs anticipated once a perceptual experiment (i.e., an action in a policy) is performed. As with scientific experiments, the greater the change in beliefs following data collection, the better the experiment. Returning to the analogy of flipping a coin and accumulating counts, this tells us something useful. If we have two coins and can choose to flip either one, we can elicit the greatest change in beliefs by flipping the coin we had flipped only five times previously rather than the coin with 100 previous flips. There is greater novelty associated with flipping the former (less famil-iar) coin. Similarly, if we have confident prior beliefs *as if* we had observed something many times, policies that interrogate these variables are associ-ated with less novelty than those about which we have less confident beliefs.

To illustrate how this works in practice, imagine we have a very myopic creature standing on a tiled floor. This creature can only see the color of the tile it is standing on and can only move one tile at a time. For any suitably large landscape with many tiles, it is very computationally expensive to represent the color of each tile as a different hidden state. However, a simpler form of model is available. If we associate hidden states only with location, and colors only with outcomes, we can efficiently represent beliefs about "what I would see if I went over there" in the **A** matrix that generates colored tiles from locations. By accumulating Dirichlet parameters (equation 7.10), our creature can optimize these beliefs on the basis of observations. We might interpret this as a form of synaptic memory as opposed to the maintenance of persistent activity in neurons representing beliefs about the color of a given tile. Given this sort of generative model, wherein all of the uncertainty is in the parameters of the likelihood distribution, it is interesting to see what happens in the absence of any preferences (i.e., when the novelty term of equation 7.11 dominates policy selection). Figure 7.11 shows a simulation of a simple environment comprising 64 black or white tiles. As each tile is visited, beliefs about the likelihood of observing black or white in that location are updated through accumulation of Dirichlet parameters. As large Dirichlet parameters preclude large belief updates, the drive to novelty resolution given by expected free energy minimization leads our simulated creature to avoid any previously visited locations.

The same principles could be applied to a range of other paradigms (e.g., if we reinterpret the path taken by our creature as a saccadic scan path, this could be applied to active visual sampling). In the domain of active vision, this has been used to simulate the kinds of visual search behavior induced by target cancellation tasks (Parr and Friston 2017a). Subsequently, evidence for the short-term plasticity required in accumulating Dirichlet parameters in this setting has been demonstrated (Parr, Mirza et al. 2019).

Just as we can extend ideas about inference to learning, it is possible to go (at least) one step further and think about structure learning: the process of not just optimizing the parameters in the model but selecting between different models with more or fewer parameters in play. Box 7.3 sets out a way of doing this that involves efficient post hoc comparisons of alternative hypothetical models. This has been used as a metaphor for sleep (Friston, Lin et al. 2017) and resting spontaneous activity (Pezzulo, Zorzi, and Corbetta 2020), where no new data are collected but the structure of the model may still be refined and simplified.

Path   Likelihood

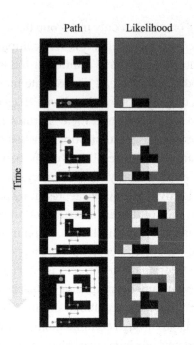

Time

Figure 7.11
Active learning is demonstrated by a synthetic creature exploring a simple world of
black and white tiles (Bruineberg et al. 2018, Kaplan and Friston 2018). *Left:* Path
taken by the creature, showing which tiles are white and which are black (dots correspond to visited locations). *Right:* A matrix of the creature and the beliefs (in terms
of normalized Dirichlet counts) the creature has about what it would see on going
to different locations. Cells in the **A** matrix are white (or black) if the creature has a
strong belief that the corresponding tile is white (or black); they are grey if the creature is uncertain about color. Crucially, these beliefs influence which path it takes via
the novelty term of the expected free energy. Those locations about which it has confident beliefs afford relatively little opportunity for uncertainty resolution, so it does
not revisit them. In other words, the phenomenon of "inhibition of return" (Posner
et al. 1985) emerges naturally from the minimization of expected free energy.

7.6   Hierarchical or Deep Inference

In the previous section, we saw one method for hierarchical extension of
the original generative model based on defining priors over the parameters
of the generative model. Figure 7.12 shows a second form of hierarchy that
speaks to the nesting of temporal scales. This generative model for hierarchical or deep inference can be conceived of as a hierarchical extension

Box 7.3

Structure learning and model reduction

The discussion in section 7.4 deals with an important, but limited, form of (parametric) learning. The next level of sophistication—in learning about the structure of the world—goes beyond the optimization of model parameters and asks whether we should expand or prune the model structure. This can be cast as a question of model comparison (Friston, Lin et al. 2017). In other words, would my free energy increase or decrease if I were to (for example) eliminate elements of a likelihood matrix? By comparing models with and without these elements, we can answer this question. However, it may be very costly to have to explicitly invert multiple models. Fortunately, an efficient method for doing this—known as Bayesian model reduction (Friston, Litvak et al. 2016; Friston, Parr, and Zeidman 2018)—is available and only requires inversion of a single full model. In a general setting, comparison between a full model and one with alternative priors (indicated by ~) can be achieved through the following formulae:

$$\Delta F = F[\tilde{P}(\theta)] - F[P(\theta)] = \ln \mathbb{E}_{Q(\theta)} \left[\frac{P(\theta)}{\tilde{P}(\theta)} \right]$$
$$\tilde{Q}(\theta) \propto \exp\left(\ln Q(\theta) + \ln \tilde{P}(\theta) - \ln P(\theta) + \Delta F \right)$$

For the Dirichlet priors used in section 7.5, this takes the form (where B is the multivariate beta function):

$$\Delta F = \ln B(\tilde{d}) - \ln B(d) + \ln B(\mathbf{d}) - \ln B(\tilde{\mathbf{d}})$$
$$\tilde{\mathbf{d}} = \mathbf{d} + \tilde{d} - d$$

This form of model reduction may be important in understanding offline model optimization, of the sort that may occur during sleep. We will briefly revisit Bayesian model reduction in chapter 8, when considering the optimization of hierarchical models with both discrete and continuous components.

of the shallow model shown in figure 7.3: it includes a series of POMDP models at the lower level that are the same as in Figure 7.3 (one example is outlined with the dashed box), contextualized by a higher-level POMDP.

Importantly, this generative model includes variables that evolve at different timescales: slower for higher levels and faster for lower levels (Friston 2008; Friston, Rosch et al. 2017; Pezzulo, Rigoli, and Friston 2018). This becomes evident if one considers that the POMDP models at level 1 evolve over three time steps, but each of these short trajectories of states and outcomes depends on a single state at the higher level (level 2) that persists throughout the entire trajectory at the lower level. In other words, for every time step from the perspective of the higher level, there are multiple (here, three) time steps for the lower level.

To gain some intuition for this separation of timescales—which underwrites deep temporal inference—it is worth thinking about a simple example of hierarchy in everyday life: reading. We draw inferences about words that combine to form sentences. Sentences combine to form paragraphs, pages, books, libraries, and so on. If we imagine that each state at the lower level of figure 7.12 is a word, each state at the higher level can be thought of as a sentence. Crucially, the duration of the sentence transcends that of any one word in the sequence.

The reading example is illustrated in more detail in figure 7.13, which is based on the example from Friston, Rosch et al. (2017), to which we refer

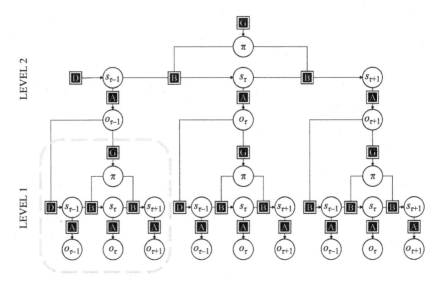

Figure 7.12
We can extend the (shallow) generative model set out in figure 7.3 so that it affords hierarchical or deep inference, which evolves over multiple timescales. The full generative model includes a slowly changing context (at level 2) that generates a series of short trajectories at the lower, faster level 1. The form of the POMDP is the same at the higher level as at the lower level (one of the POMDPs is outlined with the dashed box). The only difference is that it is stretched out in time (horizontally) and that the outcomes it generates are not directly observed. Instead, they form empirical priors for the lower level, which generates observable outcomes.

for more details. The model is structured as in figure 7.12 and represents sentences (at the higher level) and words (at the lower level) drawn from a very simple language. This language comprises three possible words (*flee, feed, wait*) that may be arranged into six possible four-word sentences. If the sentence is "flee, wait, feed, wait," the higher level predicts the word *flee* for the first of the lower-level POMDPs, *wait* for the second, and so on. At the lower level, we start with an empirical prior (**D**) based on the higher level, which tells us which words are most plausible. For example, if we started with a uniform distribution over the sentences shown in the upper panel of figure 7.13, we see that the first word is *wait* in two-thirds of the sentences and *flee* in the other third. This means that at the first time step of the first low-level POMDP, our **D**-vector should ascribe these probabilities to these words.

The words at the lower level then generate observations, visual inputs based on which part of the word is currently foveated. Much as in the example of figure 7.9, the POMDP allows for selection of different foveal targets to accumulate evidence for or against each hypothetical word. This appeals to the same expected free energy minimizing processes outlined above; therefore, we will not detail the specific foveations made here, but we note that with each time step at the lower level, there is an increase in confidence about the word in play. In the sequence shown in figure 7.13, we see that evidence is accumulated for the word *flee* at the lower level over the first few time steps (over the fast scale, $\tau^{(1)}$). This inference is propagated back up to the higher level, where it provides evidence for the first and fourth sentences (each of which start with this word). Over subsequent time steps the evidence accumulated at the lower level is consistent with both sentences. At the fourth step (at the slow scale, $\tau^{(2)}$), we would predict *wait* under the first sentence and *flee* under the second. On inferring *wait* at the fast timescale, the first sentence is inferred at the slow scale. At the final step, the simulation selects the correct sentence and is rewarded with correct feedback. The resulting belief updating is seen in the LFP plot in the lower part of figure 7.12.

Deep temporal models of this sort have been used to simulate reading (Friston, Rosch et al. 2017), delay-period working memory tasks (Parr and Friston 2017c), and computation of empirical priors for visual inference (Parr, Benrimoh et al. 2018). In addition, they have been leveraged in theoretical accounts of motivation and control (Pezzulo, Rigoli, and Friston 2018). In

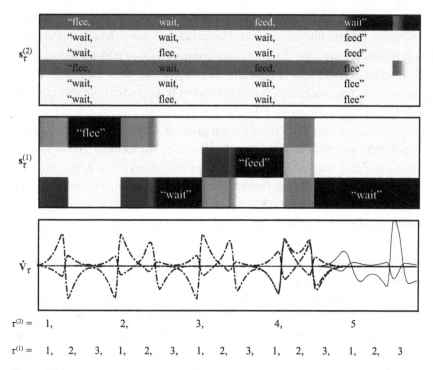

Figure 7.13
Belief updating occurs over multiple timescales by inverting a simulated hierarchical inference model. This relies on a generative model with a separation of timescales (shown as a slow timescale—$\tau^{(2)}$—and a fast timescale—$\tau^{(1)}$). Belief updating at the higher level ($s^{(2)}$), representing sentences, is slower than at the lower level ($s^{(1)}$), representing words. *Lower panel:* LFPs, i.e., the rate of change of the log expectations—which is proportional to the prediction errors (ε) shown in previous figures.

principle, these models can be extended to an arbitrary number of levels, accounting for a deeply structured world with dynamics that play out over many different temporal scales.

We can draw an interesting parallel between hierarchical models of the sort in figures 7.12 and 7.13 and learning models of the sort in figure 7.10. Learning models can be considered hierarchical generative models, which highlight a separation of timescales between faster inferential dynamics (updates of beliefs about states) and slower learning dynamics (updates of beliefs about parameters). The models shown in figures 7.10 and 7.12 may be also combined to arbitrary levels of complexity, wherein the relationships between variables on different levels may themselves be learned. This permits designing increasingly sophisticated generative models that address systems-level cognitive and neurobiological questions.

7.7 Summary

In this chapter, we saw some of the ways that discrete-time generative models may be constructed to address a range of cognitive and neurobiological problems, such as perceptual inference, decision-making and planning, balancing exploration and exploitation, parametric and structure learning, and novelty seeking. This is far from an exhaustive summary of applications of discrete models in Active Inference, but it serves to illustrate the key principles of this sort of modeling. The models outlined above may be combined hierarchically, with added priors over parameters, and with context-sensitive priors for policies or preferences. Importantly, inference using both simple and more complex generative models can always proceed through free energy minimization, which illustrates the generality of the approach. The fact that different aspects of Active Inference become apparent under distinct generative models (e.g., novelty seeking with priors over model parameters) opens up the possibility of exploring an open-ended set of cognitive and biological problems by designing the appropriate generative models.

8 Active Inference in Continuous Time

Everything flows, nothing stands still.
—Heraclitus, 501 BC

8.1 Introduction

This chapter complements chapter 7 by continuing our discussion of how to build a generative model. Our focus here is on continuous state-space models, which are well suited for modeling the physical fluctuations impinging on sensory receptors and for the continuous motion of the effectors (e.g., muscles) we use to change the world around us. There are many applications of these models. In this chapter, we set out the principles behind their use. We highlight the kinds of model used in motor control and the dynamical systems that play a role in such models, and we touch on the concept of generalized synchrony. Finally, we discuss the reconciliation of discrete and continuous generative models.

8.2 Movement Control

As we saw in chapter 4, the generative model that underwrites active inference in continuous time may be written as a pair of stochastic equations that determine how states (x) generate data (y) and how states evolve over time depending on some static variable (v):

$$y = g(x) + \omega_y$$
$$\dot{x} = f(x,v) + \omega_x \tag{8.1}$$

These equations and the precision associated with the fluctuations (ω) determine the model used to draw inferences about the causes of sensations.

Note that action is absent from equation 8.1. This is because (as outlined in chapter 6), action is part of the generative process, not the generative model. The generative model only deals with those variables that are directly influenced by states external to a Markov blanket. If we were to write down the dynamics of the real world (i.e., the generative *process*), we would have to include action (u):

$$y = \mathbf{g}(x) + \omega_y$$
$$\dot{x} = \mathbf{f}(x, u) + \omega_x \tag{8.2}$$

Note that the functions g and f (and the precisions of ω) used to define the generative model (equation 8.1) are not necessarily the same as those used to define the generative process (equation 8.2). As we saw in chapters 2–4, actions change sensory data such that free energy is minimized. This means we do not need to explicitly write down the dynamics of action in the generative model—they emerge from the choices made for the terms in equation 8.1. To gain some intuition for this, we start with a very simple sort of generative model:

$$g(x) = x$$
$$f(x, v) = v - x \tag{8.3}$$

Equation 8.3 says that the hidden state represents the expected value for the data and that it has dynamics consistent with a simple (i.e., point) attractor. By *attractor*, we mean that when x is less than v, the expected rate of change of x is positive, and vice versa. This means that x will always flow toward v (i.e., v is an *attracting* or *fixed* point). To generate data, we define a simple generative process:

$$\mathbf{g}(x) = x$$
$$\mathbf{f}(x, u) = u \tag{8.4}$$

On minimizing free energy, this means that action will change to fulfill the predictions of equation 8.3. If μ is the expected value of x, this means the action that minimizes the difference between the predicted data ($g(\mu)$) and the observed data (y) is to set u equal to $v - \mu$. This is an expression of the "equilibrium point hypothesis" (Feldman and Levin 2009), which treats motor control as enacted by reflex arcs that simply draw limbs toward equilibrium points set by descending motor signals. Under Active Inference, these signals are predictions—specifically, *proprioceptive* predictions about, for example, the expected position of limbs or eyes (Adams, Shipp, and

Friston 2013). Therefore, movement control results from the fulfillment of (proprioceptive) predictions by action, as schematically illustrated in figure 8.1. Note that this scheme does not require specification of "inverse models" (i.e., mappings from desired consequences to the motor commands to reach them) that are widely used in other formulations of motor control (Wolpert and Kawato 1998).

The expression in equation 8.3 is the simplest sort of attractor system we might employ in a generative model. However, it is too simple in many settings, where more realistic Newtonian dynamics apply. A more sophisticated model recognizes that forces—generated by muscles—change the velocity (i.e., induce an acceleration), not the position. Equation 8.5 sets this out explicitly with x_1 as the position and x_2 as the velocity:

$$f(x,v) = \begin{bmatrix} x_2 \\ \frac{\kappa}{m}(v - x_1) \end{bmatrix} \tag{8.5}$$

This expression is equivalent to the dynamics of a spring obeying Hooke's law. The rate of change of the position (first element) is simply the velocity. The rate of change of the velocity (second element) is proportional to the distance between the current position and the point v, with the constant of proportionality: a ratio between the mass of the object (m) and a (spring) constant (κ). Multiplying both sides by the mass, we have the force[1] generated by a spring ($\kappa(v-x_1)$) attached to the points v and x_1 equal to the mass multiplied by the rate of change of the velocity. This is just Newton's second law. In other words, we can write down a generative model that predicts the dynamics that would unfold if there were a spring drawing a limb to a desired location. By predicting the (proprioceptive) data consequent on this Newtonian mechanics, we can enact the movement that fulfills these predictions.

8.3 Dynamical Systems

As outlined in section 8.2, continuous-time formulations of Active Inference are well suited to characterization of movements. More generally, they are appropriate in specifying generative models of nonlinear dynamical systems wherein discretization of time and space is inefficient. The simplest form of dynamical system is the attractor of equation 8.3, but much richer behavior can be developed from more complex systems. In the limited space of this book, we cannot do justice to the large body of

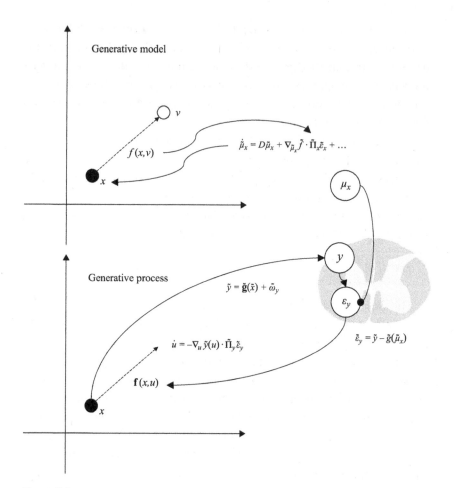

Figure 8.1

Spinal reflexes, illustrating the distinction between a generative process (out there in the world) and a generative model in the setting of action generation. The model assumes that the position (x) of a limb (or hand or other body part) is drawn toward some point (v). The dashed arrow in the upper plot shows this belief. Beliefs about $x(\mu_x)$ may be substituted in place of x and used to update beliefs about its rate of change. The resulting μ_x is then used to predict sensory data (y) via the g function in the generative model. Sensory data are actually generated by the generative process via the function **g**, which takes the "real" value of x as its argument. The error (ε_y) then drives changes in action (u) such that the error is resolved. This resolution happens through the generative model, as the action determines the rate of change of x via **f**. This causes x to move to the location in space that generates data y consistent with the prediction ($g(\mu_x)$), setting ε_y and therefore the rate of change of a to zero.

Box 8.1

Precision, attention, and sensory attenuation

We addressed the importance of precision in chapter 7, but it is worth recapping its role in continuous-time systems. In many ways, this concept is more naturally addressed in this setting, as the Π variable appears as a direct consequence of the Laplace approximation. This acts directly as a multiplicative gain in the inferential dynamics (see figure 8.1), with different precisions weighting alternative influences over belief updating.

The interpretation of precision as a synaptic gain connects it to several important aspects of neurobiology. From an empirical point of view, higher precision implies more vigorous belief updating of the sort that might be measured in electrophysiological research as a large amplitude-evoked response with an early peak or in single-cell recordings as a multiplicative effect on neuronal firing rates in response to a stimulus placed in that cell's receptive field. These findings are often associated with attentional processing, where one sensory channel (or subset of channels) is favored above others. From the perspective of active inference, *precision* and *attention* are synonyms. The former has been used to reproduce a range of attentional phenomena in silico, including the Posner paradigm (Feldman and Friston 2010). Specifically, using a cue to predict the precision of sensory input from one of two locations reproduces the empirical finding that responses to stimuli in the cued location are faster than those appearing in the alternative location.

A second important aspect of precision control is its role in movement generation. To understand this, it is worth thinking about what happens in the absence of this control. Imagine, first, that sensory data are predicted with high precision. The messages from these data are therefore afforded high synaptic gain and lead to veracious inferences about the position of some body part. The problem with this is the equivalence between motor commands and predictions under Active Inference. An accurate belief that "I am not moving" cannot be used to predict the sensory consequences of movement, vital for the initiation of that movement. With high precision sensory input, the belief that "I am moving" is immediately corrected in the face of evidence to the contrary; hence no movement is executed. This tells us something important: In order to generate movement, we must be able to ignore the sensory consequences of that movement to form the (initially false) belief that "I am moving." Once this belief is established, the proprioceptive (and other sensory) consequences of that movement may be predicted and enacted through the mechanisms outlined in figure 8.1. This process of ignoring evidence to the contrary is known as "sensory attenuation" and represents the decrease in precision required for a movement to take place (Brown, Adams et al. 2013; Pezzulo 2013; Seth 2013; Pezzulo, Rigoli, and Friston 2015; Seth and Friston

Box 8.1 (continued)

> 2016; Allen et al. 2019). Clearly it is useful between movements to restore this precision, to draw the appropriate inferences from sensory input. This implies a cyclical process of attenuating and moving (e.g., the cyclical suppression of visual input during saccades, then a suppression of saccades). Predicating movement on the suspension of attention has close relationships with an *ideomotor theory* that originated in the nineteenth century to explain movements induced under hypnosis.

work developing models with more complex dynamical systems (but see table 8.1 for some of the key advances). Instead, we focus on a few of the principles needed to understand these systems. In this section, we briefly overview two dynamical systems used in formulating generative models of this sort: Lotka-Volterra dynamics and Lorenz systems. The former may be used in characterizations of systems with a sequential aspect to their dynamics, while the latter represent chaotic systems.

Lotka-Volterra dynamics inherit from characterizations of predator-prey dynamics in ecology. While they have since found application in numerous disciplines, predator-prey systems remain a useful example to provide some intuition about their workings. When the predator population is small, the prey may increase their numbers to become a relatively large population. This provides additional food for the predators, whose population size then grows. Increased predation causes a decrease in the number of prey species and therefore a decrease in the number of predators. From here, the cycle continues. This gives an oscillatory pattern whereby the prey population size peaks, then the predators', then the preys' again, and so on. By generalizing this to more than two populations (e.g., carnivore, herbivore, and plant populations), we can generate a sequence of peaks. Figure 8.2 illustrates generalized Lotka-Volterra dynamics with three populations, which obey dynamics of the following form:

$$f(x, v) = x \circ (v + \mathbf{A}x) \tag{8.6}$$

Here, x is a vector as before. The $\circ$ symbol means an elementwise product. Intrinsic birth and death rates are given by the vector v, and $\mathbf{A}$ is a matrix whose elements are positive if the species indexed by the column prey on those indexed by the row and negative if the relationship is inverted.

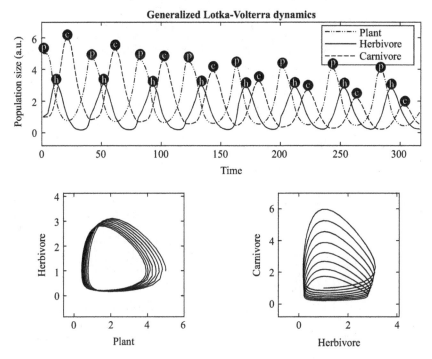

Figure 8.2

Generalized sequential dynamics emerging from Lotka-Volterra systems provide an important point of connection with the discrete sequential dynamics assumed in chapter 7. These dynamics can be applied to a range of systems but are framed here in terms of predator-prey relationships for ease of interpretation. *Top:* Population changes over time. The population size is expressed in terms of arbitrary units (a.u.). The peaks are labeled on the basis of which species has the greatest population at that point. The repeated pattern of *p, h, c* can be seen as a sequence of three (not necessarily evenly spaced) discrete time steps. *Bottom:* Trajectories emphasizing the (approximately) periodic pattern that each follows.

Figure 8.2 makes clear that having a generative model that incorporates Lotka-Volterra dynamics allows for temporal sequencing (Huerta and Rabinovich 2004)—depending on the current highest peak. Each line can be thought of as representing a hidden state, in place of a species. Figure 8.3 highlights two important examples in which these dynamics have been exploited to generate behavior One is a hierarchical model that uses the sequential dynamics afforded by a Lotka-Volterra system to time the response of an eye blink relative to a conditioned stimulus (Friston and Herreros 2016). The

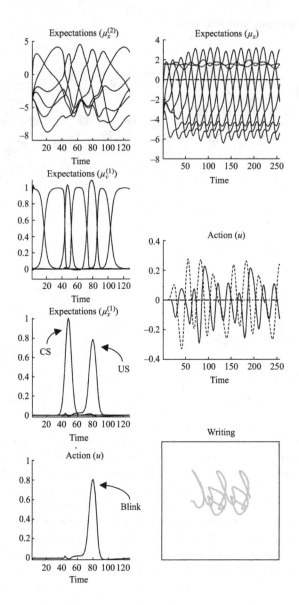

Figure 8.3

Two applications of generalized sequential Lotka-Volterra dynamics in Active Inference. *Left:* Eyeblink conditioning used to empirically investigate cerebellar function (Friston and Herreros 2016). Starting at the highest level of the column, the expected states show the same kind of sequential pattern as in figure 8.2. This passes down to the next level to predict sequential hidden causes; the various peaks here predict states at the next level down, where the first peak is the conditioned stimulus (CS) and the second is the unconditioned stimulus (US). Finally, the predicted US induces action—a blink. *Right:* Sequential peaks using an attracting point as in equation 8.5 but selecting the specific attractor on the basis of which population of a Lotka-Volterra system is currently highest; this leads to a sequential visiting of each point, giving rise to a form of handwriting (Friston, Mattout, and Kilner 2011).

paradigm is based on those used in the investigation of cerebellar function. An unconditioned stimulus (a puff of air directed toward an animal's eye) elicits a response (blinking). A conditioned stimulus (an auditory tone) may be played prior to the unconditioned stimulus on multiple occasions. By learning (see box 8.2) the number of peaks in the Lotka-Volterra dynamics that separate the conditioned stimulus from the unconditioned stimulus, the animal learns to preempt the air puff and time the appropriate blink. This is a form of temporal learning, since the number of peaks provides an implicit estimate of the length of the temporal interval from the conditioned to the unconditioned stimuli. In the second example in figure 8.3, each sequential peak is associated with an alternative attracting point that drives movements to a series of attracting points arranged to suggest handwriting (Friston, Mattout, and Kilner 2011). As the two examples illustrate,

Box 8.2

Learning in continuous models

As discussed in chapter 7, learning is the process of optimizing beliefs about the parameters (θ) of a generative model. In the continuous-time domain, this means accumulating evidence over time. This works as if we treat data in a series of infinitesimally small time-intervals as obeying i.i.d. (independent and identically distributed) assumptions and formulate a generative model that generates observations from (time-invariant) parameters:

$$\ln p(\tilde{y}, \theta) = \ln p(\theta) + \int \ln p(y(t) \mid \theta) dt$$

$$\approx \ln p(\theta) - \int F[y(t) \mid \theta] dt$$

This may be used to formulate a functional (S) that plays the role of a free energy for parameters using the time integral of the free energy conditioned on parameters. Using a Laplace approximation, we get the following, wherein α acts to accumulate free energy gradients (i.e., evidence gradients):

$$S(\theta) = E_{q(\theta)}[\ln q(\theta) + \int F[y(t) \mid \theta] dt - \ln p(\theta)]$$

$$\approx \int F[y(t) \mid \mu_\theta] dt - \ln p(\mu_\theta)$$

$$\dot{\mu}_\theta = -\partial_{\mu_\theta} S(\mu_\theta)$$

$$= \partial_{\mu_\theta} \ln p(\mu_\theta) - \int \partial_{\mu_\theta} F[y(t) \mid \mu_\theta] dt$$

$$= \partial_{\mu_\theta} \ln p(\mu_\theta) - \alpha$$

$$\dot{\alpha} = \partial_{\mu_\theta} F[y(t) \mid \mu_\theta]$$

generalized Lotka-Volterra systems afford useful models of sequential dynamics using a continuous dynamical system.

The POMDP formulation of chapter 7 has largely superseded the use of generalized Lotka-Volterra systems in Active Inference applications. However, it is useful to bear this kind of dynamic in mind as a plausible continuous system that might underwrite the discrete sequential dynamics of chapter 7. In addition, Lotka-Volterra systems make explicit the distinction between representations of sequences involved in temporally deep planning and representations of rates of change in generalized coordinates of motion (see chapter 4). Each has its place but deals with different sorts of problems.

The second sort of dynamical system that has found widespread application in active inference research is the Lorenz system:

$$\dot{x} = \begin{bmatrix} \sigma(x_2 - x_1) \\ x_1(\rho - x_3) - x_2 \\ x_1 x_2 - \beta x_3 \end{bmatrix} \tag{8.7}$$

The parameters are known as the Prandtl number (σ), the Rayleigh number (ρ), and a constant (β) that relates to the physics of the system. Depending on the values these take, the system may behave in very different ways. Lorenz attractors were initially formulated to account for atmospheric convection dynamics. Their itinerant (wandering) behavior has prompted their use in generative models to simulate challenging inference problems. An important example of this is in the simulation of birdsong, which we unpack in the next section. These systems have also been used to simulate simple physical systems and to investigate the conditions under which their behavior starts to appear sentient. Figure 8.4 shows how the Lorenz system behaves under example parameter settings.

8.4 Generalized Synchrony

As mentioned above, a key example application of continuous state-space models is in a series of studies based on synthetic birdsong (Friston and Frith 2015b). An important aspect of these studies looks at communication and multi-agent inference problems. The idea here rests on the capacity of a creature to synchronize its internal states with something out there in the world (i.e., inference). When what is out there is another creature with a similar model, this synchronization means the internal states of one

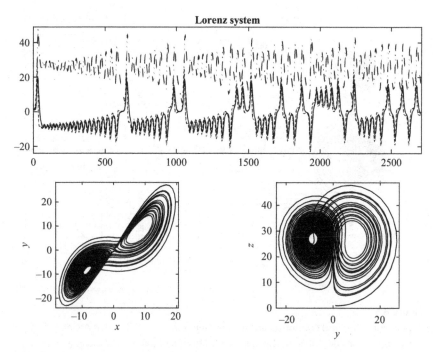

Figure 8.4
Behavior of a Lorenz system attractor (using the same format as figure 8.2), showing how this 3-dimensional system evolves. Characteristically, it appears chaotic and unpredictable, spending some of its time orbiting one part of space before switching to another orbit. This itinerancy and apparent autonomy make this interesting system well suited to inclusion in models of biological phenomena.

creature should come to resemble the internal states of the other: a primitive kind of *theory of mind*.

Figure 8.5 shows the kind of generative model used to simulate songbirds. In this hierarchical model, high-level states (level 2) evolve according to a slow Lorenz system. One dimension of this system is then used to parameterize the Rayleigh number of a faster Lorenz system at the lower level (level 1). The lower-level variables then map to sensory (sonographic) data. Analogous to figure 8.1, the generative process additionally includes action; here, instead of moving a limb, actions influence the larynx, such that the sonographic data may be influenced by the bird. As before, action is generated to resolve prediction error. This means that if a bird hears the song it is predicting, there is no need to generate it itself. However, if it predicts a song that is not heard, it must start singing to resolve any error.

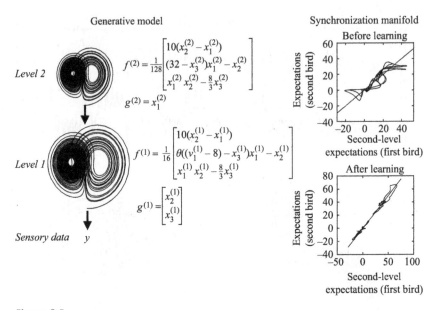

Figure 8.5

Synchronization and communication. *Left:* Generative model underwriting the bird-song simulations described in the main text. This is a hierarchical model, with Lorenz attractors at each level. *Right:* Synchronization manifolds of expectations at the second level for two birds before and after they have learned about one another. After learning the parameters of each other's generative models, the two bird's joint trajectory is confined to an (almost) 1-dimensional subspace, indicating synchronization.

This dynamic becomes more interesting when there are two birds in play, with similarly structured generative models. As long as one bird is singing, the other does not need to, as there is no error to resolve. However, if one bird stops singing, the other needs to continue the same song. This leads to a form of turn taking, sometimes phrased as "singing from the same hymn sheet," with each bird contributing sections of the same song. What leads to this turn taking? Why doesn't one bird continue singing the whole song to its conspecific? The answer relates to the issue of sensory attenuation (see box 8.1), as acting to generate birdsong requires a reduction in the precision of predictions about the consequences of action. Just as in saccadic eye movements, this implies alternation between attention to sensory (visual or auditory) data and attenuation during (saccadic or vocal) action designed to change those data. When there are two agents involved, this leads to an alternation between listening to the other and singing—a simple form of conversation.

For this synthetic conversation to work, it is essential that the two birds synchronize with one another and know where they are in the song (or conversational trajectory). This implies that the inferences about hidden states in the generative model should be aligned between the birds. On the upper right of figure 8.5, we show a synchronization manifold of two birds who have not yet optimized their generative models in relation to one another; this plots a trajectory of the beliefs each bird has about the higher-level hidden states. Synchronization implies that when one bird infers a specific hidden state value, the other bird should infer the same; therefore, we would expect the trajectory to stay fixed to the $x = y$ line (technically called *identical synchronization of chaos*). Fluctuations around this line imply imperfect synchronization, as this plot shows. After exposure to one another and learning the parameters of each other's generative models (see box 8.2), the synchronization is nearly perfect (lower-right plot). The implication is that each bird has learned about the other and is able to infer what is going on in the other's head. In short, they have learned to share the same narrative and "sing from the same hymn sheet."

A more general form of synchronization does not require synchronization along the $x = y$ line. In generalized synchronization, the joint behavior occupies a lower, 1-dimensional space than the higher, 2-dimensional space that could be occupied by this behavior. However, this low-dimensional space (the synchronization manifold) may be curved or have some other shape; this is analogous to the 2-dimensional space we occupy on the surface of the planet, despite the surface being curved into a 3-dimensional sphere. In addition to its central role in social behavior, generalized synchronization—occupancy of a low-dimensional region of a high-dimensional joint space—is very important in characterizations of biological systems as engaging in inference (generalized synchrony between internal and external states). While we do not have the space to unpack this extensive subject here, the inferential perspective speaks to the failure of generalized synchrony associated with neuropsychiatric syndromes like autism. This kind of synchrony is important not only in continuous-time models but also in POMDP models of linguistic communication between multiple agents (Friston, Parr et al. 2020).

8.5   Hybrid (Discrete and Continuous) Models

As we have seen in this and the previous chapter, discrete and continuous models both have important applications in Active Inference. While many settings call for one or the other, a more holistic perspective acknowledges that both are likely in play. This means we need a way to combine these generative models so that a single model includes both continuous and discrete variables (Friston, Parr, and de Vries 2017). Such hybrid or mixed models allow inferences about sequential action plans and translations of these decisions into movements through a continuous model. Figure 8.6 shows the form of these models, with a POMDP at the higher level, behaving as described in chapter 7, which generates a continuous model of the

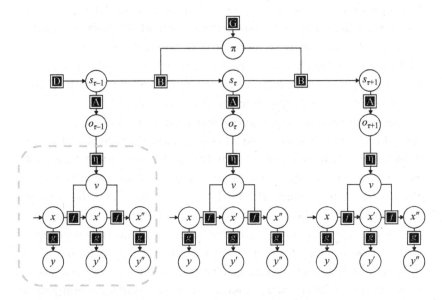

Figure 8.6
Mixed generative models in the form of a hierarchical model much like that in figure 7.12. However, there is an important difference between the form of the model at the higher level and that at the lower level. While the lower-level model (one example is highlighted by the dashed box) is the same form as the other models considered in this chapter—i.e., it is framed in terms of continuous states and continuous time and uses generalized coordinates of motion—the higher-level model is a POMDP model of the sort we saw throughout chapter 7—i.e., it is framed in terms of discrete states and times. Effectively, this means we can select (at regular time intervals) between alternative segments of a continuous trajectory.

sort addressed in this chapter at each discrete time step. This decomposes continuous time into a discrete sequence of short continuous trajectories.

To translate from the outcomes of the discrete level to the continuous level, we need to associate each alternative outcome with a point in some continuous space. To develop intuition for this idea, we consider the example of a delay-period oculomotor task—often used in primate research; see figure 8.7. This task involves three stages. First, a target appears in one of (for example) four possible locations, while a monkey maintains fixation on a central fixation cross. Next, the target disappears and must be remembered during a delay period. Finally, a signal is given for the monkey to make a saccade, at which point it must look at the location where the original target appeared. To complete this task, the monkey must be able to draw inferences about sequences (which stage of the task is currently in play) and to infer which of four locations to aim for. These are categorical inference problems suited to a POMDP formulation. However, once the appropriate location has been selected, the monkey must perform the eye movement that brings its fovea to the (continuous) coordinates of the target location.

Figure 8.7 illustrates the functioning of a mixed generative model that solves this problem. In the top panel, the model's higher level makes categorical decisions: it computes the posterior beliefs about four discrete target locations at four time periods. In the middle and bottom plots, the model's lower level computes continuous behavioral trajectories (eye movements) resulting from discrete inferences at the higher level.

Transforming decisions about discrete target locations into continuous eye movements requires each discrete target location (o) to be associated with a distribution over continuous hidden causes (v), which identifies the target coordinates. The prior over target coordinates can then be computed by taking the Bayesian model average over these locations, weighted by the inferences at the POMDP level:

$$P(\tilde{v}\,|\,o_\tau) = \mathcal{N}(\tilde{\eta}_{o_\tau}, \tilde{\Pi}_v)$$
$$P(\tilde{v}) \approx \mathcal{N}(\tilde{\boldsymbol{\eta}}, \tilde{\Pi}_v) \tag{8.8}$$
$$\tilde{\boldsymbol{\eta}} = \mathbb{E}_{Q(o_\tau)}[\tilde{\eta}] = \mathbf{o}_\tau \cdot \tilde{\eta}$$

To infer which discrete target best explains continuous data, we need to be able to compute the evidence associated with each hypothetical target—which is a function of observed continuous data. This exemplifies the fact that mixed models require reciprocal interactions between higher and lower hierarchical levels. As we see in figure 8.7, this facilitates the formation of

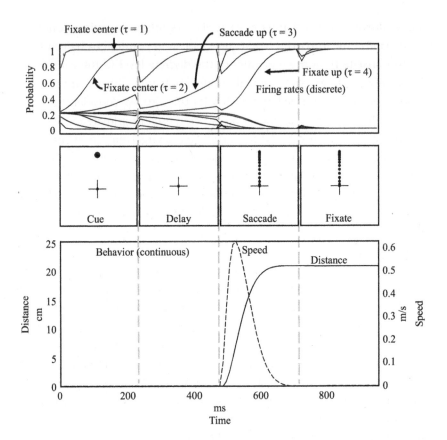

Figure 8.7
Transforming decisions into movement using mixed or hybrid models (Parr and Friston 2019b). This simple example uses the oculomotor delay-period paradigm outlined in the main text. *Top:* Neuronal firing rates representing posterior beliefs about the target location. The target may be in four different locations, and there are four time steps in this synthetic experiment, so there are 16 neural populations representing each of these combinations. The lines corresponding to the final inferred state of affairs are annotated explicitly. Note the belief updating at the first time step (from 0 to 250 ms), when the target initially appears, and at the third time step (from 500 to 750 ms), when the agent observes itself performing a saccade to that location. *Middle:* Behavior (i.e., a saccade from the center location to the target location). During the first quarter of the experiment, the target is visible in the upper location. Next, there is a delay period, during which fixation is maintained. Then a saccade is performed to the correct location. Finally, there is a period during which fixation is maintained at the target location. *Bottom:* Continuous behavioral trajectories that result from the discrete inferences of the top plot.

Box 8.3

Mixture models and clustering

The issue of combining categorical and continuous generative models outside of Active Inference has primarily been framed through the lens of clustering. Here, the aim is to assign each (continuous) data point to a (discrete) cluster. A range of algorithms have been employed to solve this problem, but most of them implicitly rely on a generative model similar to that used here. This is a *mixture of Gaussians* (aka a *Gaussian mixture model*):

$$P(\tilde{y}, \tilde{s}, \mathbf{D}, \eta, \Pi) = P(\mathbf{D})P(\eta)P(\Pi)\prod_i P(s_i \mid \mathbf{D})P(y_i \mid s_i, \eta, \Pi)$$

$$P(\mathbf{D}) = Dir(d)$$
$$P(s_i \mid \mathbf{D}) = Cat(\mathbf{D})$$
$$P(y_i \mid s_i, \eta, \Pi) = \mathcal{N}(\eta_{s_i}, \Pi_{s_i})$$

The problem in clustering approaches is to infer the mean and precision (η and Π, respectively) of each cluster and the posterior probability that each data-point (y_i) belongs to a given cluster $P(s_i \mid y_i)$. For our purposes (as described in section 8.5), we assume precise (delta-function) priors for η and Π and calculate $Q(s_i) \approx P(s_i \mid y_i)$ via Bayesian model reduction (see box 7.3).

---

beliefs about where to perform saccades and the execution of those saccades at the appropriate times. At the first discrete time step (up to 250 ms on the continuous scale), the monkey is able to infer with some confidence that its eyes are centered on the fixation cross during the first time step, that it will maintain this fixation at the second time step (up to 500 ms), and that the most likely course of action after this will result in foveating the upper location. This can be seen in the discrete firing rates (top plot). This translates into the continuous behavior which, when implemented, increases the confidence in beliefs about the discrete states (note the increase in probability for an upward saccade at the third time step once the continuous data become available between 500 and 750 ms). This simple example, based on experimental cognitive research, illustrates the basic principles of translation between discrete action plans and their continuous implementation.

8.6 Summary

In this chapter, we have overviewed the applications of continuous-time generative models under Active Inference. This is a huge topic, and much has been left out (see table 8.1 for further reading). However, the broad

Table 8.1
Key advances in continuous-time models

Application	Sources	Notes
Synthetic birdsong	Friston and Frith 2015a Friston and Frith 2015b Isomura, Parr, and Friston 2019	This series of papers deals with communication and the interaction between synthetic agents, a simulated pair (or group) of songbirds singing to one another. The studies unpack phenomena from generalized synchrony to perceptual inference to sensory attenuation.
Oculomotor delays	Perrinet, Adams, and Friston 2014	By taking advantage of beliefs about the near past and future implicit in models formulated in generalized coordinates of motion, it is possible to account for sensorimotor delays through projections a short way into the future or past.
Conditioned reflexes	Friston and Herreros 2016	Using a model based on a Lotka-Volterra system, the temporal relationship between a conditioned and unconditioned stimulus is learned and used to generate an anticipatory blink.
Smooth pursuit eye movements	Adams, Perrinet, and Friston 2012	This work looks at the role of smooth pursuit eye movements, following a visual target. It aims to reproduce differences between neurotypical and schizophrenic individuals in response to pursuit with and without visual occlusion.
Psychosis	Adams, Stephan et al. 2013	Building on the songbird and smooth pursuit models, this research looks at how false, psychotic inference may arise from suboptimal prior beliefs.
Illusions	Brown and Friston 2012 Brown, Adams et al. 2013	Illusions offer a useful tool to reveal the prior beliefs our brains appeal to in the presence of uncertain or ambiguous sensory input. These papers take several examples of common illusions and demonstrate the optimality of illusory inferences under certain prior beliefs.

Table 8.1
(continued)

Application	Sources	Notes
Saccades	Friston, Adams et al. 2012 Donnarumma et al. 2017 Parr and Friston 2018a	Like the smooth pursuit simulations, these papers consider eye movement control. However, here the eyes do not simply follow a target but must move to one of several possible target locations. They deal with the generative models we need to be able to do this and (once the models have been specified) the emergent architectures and physiology.
Action observation	Friston, Mattout, and Kilner 2011	This work considers the role of the mirror neuron system and formalizes the idea that generative models of our own actions can also be put to use in modeling, and replicating, behavior observed in others.
Attention	Feldman and Friston 2010 Kanai et al. 2015	Through predicting precision, we implicitly select the data that we believe is most informative. This work highlights how implementations of this idea reproduce classical psychophysical findings in the Posner paradigm and figure-ground discrimination tasks.
Hybrid models	Friston, Parr, and de Vries 2017 Parr and Friston 2018c Parr and Friston 2019b	These models make use of discrete POMDP models in combination with predictive coding schemes. Most current examples of this modeling are framed in terms of visual search behavior or oculomotor control. These require selecting where to look and then implementing the process of looking there.
Self-organization	Friston 2013 Friston, Levin et al. 2015 Palacios et al. 2020	This line of research is based on the idea that groups of cells can organize into a predefined structure when each cell has the same implicit generative model of that structure. Specifically, they must know what sort of sensory input they would predict if they were a particular kind of cell.

concepts outlined here provide a foundation from which these models may be further explored. Specifically, we have considered movement generation in terms of the fulfillment of predictions. This greatly simplifies the treatment of motor control problems, as we do not need any additional machinery or inverse models—just spinal or brain stem reflex arcs. We highlighted the role of precision and sensory attenuation in motor control of this sort. Given that a key advantage of continuous schemes is to articulate generative models in terms of dynamical systems, we outlined two ubiquitous types of dynamical system that have found widespread application in Active Inference research. Generalized Lotka-Volterra systems act to provide temporal sequencing in a continuous context, while Lorenz attractors may be used to generate rich simulations, including synthetic birdsong. Next we considered the concept of generalized synchrony. Synchronization of the internal states of a system to external states forms the basis of inferential treatments of brain function and is crucial in accounts of social systems—where external states largely comprise conspecifics (i.e., creatures like me). Finally, we set out the unification of the discrete and continuous models of chapters 7 and 8, bringing together the expected free energy minimizing (exploitative and explorative) dynamics of POMDP formulations, the enaction of the behaviors these mandate through continuous processes, and the reciprocal message passing that mediates this interaction. In short, this takes us from decisions to movements—and back again.

9 Model-Based Data Analysis

Just because we have the best hammer does not mean that every problem is a nail.
—Barack Obama

9.1 Introduction

Ultimately, the models described in this book are only useful if they can answer scientific questions. In this chapter, we focus on the ways in which Active Inference can be applied in understanding empirical data. The central idea is that we, as scientists, can appeal to the same maths as we have assumed the brain uses in previous chapters. Our general goal is to recover the parameters of the generative model that a subject's brain uses to produce behavior—the *subjective* model. For this, we can use our own generative model (of how the subjective model produces behavior)—the *objective* model. We can invert our objective model on the basis of the behavior we observe to draw inferences about the parameters of the subjective generative model. This meta-Bayesian inference affords the opportunity to test hypotheses about the model we assume the brain uses and to phenotype individuals on the basis of the prior beliefs they would have to hold for their behavior to be Bayes optimal. Belief-based computational phenotyping of this sort holds promise in the emerging fields of computational psychiatry, neuropsychology, and neurology.

9.2 Meta-Bayesian Methods

This chapter deals with the utility of Active Inference formulations in analyzing data from behavioral experiments. This goes beyond the proof-of-principle simulations we have seen in previous chapters and instead exploits Active Inference in answering scientific questions. We have seen already that a subject's generative model is the key determinant of behavior under Active Inference. This implies that hypotheses about the causes of empirical behavioral measurements must be framed in terms of the alternative generative models used to select those actions. Our challenge, then, is to fit an Active Inference scheme to observed data by manipulating the parameters (i.e., prior beliefs) of the generative model.

Broadly speaking, there are two (related) reasons for fitting a computational model to observed behavior. The first is to estimate parameters of interest from that model that best explain the behavior of a specific subject or group of subjects. This is useful in characterizing subjective behavior in terms of the computations that generate it, a process known as *computational phenotyping* (Montague et al. 2012, Schwartenbeck and Friston 2016, Friston 2017). Computational phenotypes may be used in combination with other measures (e.g., to establish links between neuroimaging findings and function) or may be used alone in forecasting behaviors in other settings (e.g., following a therapeutic intervention).

The second reason is to compare alternative hypotheses, expressed as models, that represent different explanations for a behavioral phenomenon (Mirza et al. 2018). These two agendas—parameter estimation and model comparison—map to one side of Bayes' theorem. Parameter estimation is the process of finding the posterior probability, under a model, of a parameter setting. Model comparison rests on finding the marginal likelihoods (i.e., evidence) for each model. To recap, Bayes' theorem is

$$\underbrace{P(u\,|\,\theta,m)}_{\text{Likelihood}}\,\underbrace{P(\theta\,|\,m)}_{\text{Prior}} = \underbrace{P(\theta\,|\,u,m)}_{\text{Posterior}}\,\underbrace{P(u\,|\,m)}_{\text{Evidence}}. \tag{9.1}$$

The right-hand side deals with the posterior probability of parameters (θ) given behavioral data (u) under a model (m) and the model evidence, and the left-hand side tells us what we need to specify for our model: we need prior beliefs about our parameters of interest and a likelihood function.

Importantly, while we appeal to the same Bayesian inference scheme as used in previous chapters, our purpose is different here. This rests on the fact that there are two inference processes going on (figure 9.1). The first is that creatures use their model of the processes generating their sensory data to draw inferences about their world (and about how to act). This has been the focus of the preceding chapters. The second is that we as objective

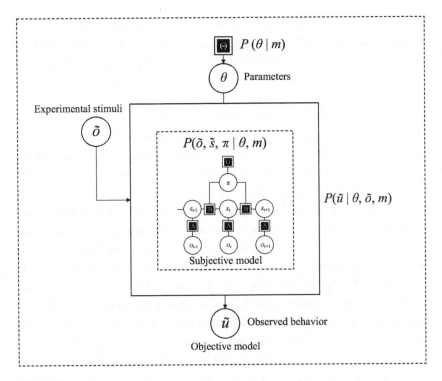

Figure 9.1

Relationship between the subjective and objective models of meta-Bayesian inference. *Inner dashed box:* Subjective model assumed to be used by an experimental subject. This could be a POMDP model as illustrated or some other form of model. The important features are that it depends on parameters (θ) whose value we do not know and that it generates sensory data (o). *Outer dashed box:* Experimenter's objective model (m) includes prior beliefs about the parameters and predicts the behavior (u) we would expect on presenting experimental stimuli (sensory data from the subjective model's perspective). Crucially, the likelihood distribution of the objective model depends on the subjective model. This means we evaluate the likelihood of parameters taking a particular value as follows. First, we incorporate the parameters in the subjective model. We then use the Active Inference schemes described in previous chapters to solve this model, presenting our experimental stimuli as sensory data, and infer a distribution over the most probable course of action. Finally, we evaluate the probability of the observed actions or choices, given this distribution. This is the likelihood of observed behavior given parameters and stimuli—i.e., the likelihood distribution in the objective model.

scientists observe the creature's behavior and seek to draw inferences about the (*subjective*) generative model it is using by inverting our own (*objective*) generative model. The implication here is that we are drawing inferences about an inferential process—sometimes referred to as "meta-Bayesian" inference (Daunizeau et al. 2010).

More formally, this approach defines the likelihood distribution in terms of the solution to an Active Inference problem. By using a given parameter setting, we can simulate behavior under Active Inference and quantify the likelihood that a series of actions would have been taken. Equipped with prior beliefs about the value of these parameters, we have a generative model of how a creature uses its generative model to produce actions. While our focus is on Active Inference (and discrete-time models specifically), the generic methods used here may be used with any arbitrary likelihood function. Other normative models of behavior (such as those used in reinforcement learning) may be substituted in place of the Active Inference models.

The following sections unpack an example of a generic inference scheme that may be used for meta-Bayesian inference (namely, variational Laplace) and the use of hierarchical models for model comparison. We then provide a simple recipe for model-based data analysis and finally review a key example of this procedure. It is important to emphasize that understanding the technical details is not required to use these methods effectively; thus, readers uninterested in these details are invited to skip sections 9.3 and 9.4.

In brief, the basic idea is to evaluate the likelihood of any observed set of choices, given the unknown parameters of interest—namely, the parameters of a subject's prior beliefs. We then combine this likelihood with our objective prior over those parameters to evaluate the posterior over the subject's priors, in the usual way. If we have several subjects, these posteriors can be combined to make inferences about group or between-subject effects, using parametric empirical Bayes (PEB). The requisite likelihood is simply the probability of sampling the observed sequence of choices, under the subject's posterior beliefs about action. These posterior beliefs depend on what the subject sees (i.e., cues or stimuli) and her prior beliefs—and are evaluated in a straightforward way by solving the appropriate Active Inference scheme. Note that we are using Bayesian procedures twice: first to evaluate the subject's posterior beliefs about action, and second to evaluate our posterior beliefs about the unknown priors that characterize the subject. We now rehearse the various parts of this meta-Bayesian procedure.

9.3   Variational Laplace

Variational Laplace is an inference scheme based on the same principles as predictive coding (Friston, Mattout et al. 2007). However, it may be used for more generic likelihood functions than those encountered earlier—which were defined as Gaussian. We will start this section with an overview of the likelihood function $\mathcal{L}(\theta)$ of interest here. This should give the probability of actions for an Active Inference scheme with a generative model with parameters set at value θ. The actions selected depend on the observations made:

$$\mathcal{L}(\theta) = \ln P(\tilde{u} \mid \theta, m, \tilde{o})$$
$$P(\tilde{u} \mid \theta, m, \tilde{o}) = \tilde{u} \cdot \sigma(\theta_\alpha \ln \tilde{u})$$
$$\tilde{u} = \pi \cdot U \tag{9.2}$$
$$\pi = \arg \min_\pi F$$

Unpacking this, the first term gives the log likelihood of an observed sequence of actions ($\tilde{u}$) as a function of parameters (θ), the model (m), and a sequence of stimuli ($\tilde{o}$) presented during a real experiment. The probability of these actions is found by using the parameters to set the prior beliefs in a POMDP model of the sort described in chapter 7. We can then solve the POMDP as described in chapters 4 and 7, forcing the simulation to take the observed action sequence and presenting it with the same experimental stimuli. As we described in preceding chapters, this involves computing the beliefs (π) a synthetic subject holds about the policy or course of action she chooses to pursue. This minimizes the free energy (F) associated with her generative model of the world. We can then take these beliefs and calculate the average probability of pursuing an action sequence. This requires us to distribute the probability for each policy over the actions implied by that policy (indexed by an array U). Finally, a softmax temperature parameter (θ_α) is applied to account for randomness (*shaky-handedness*) in behavior not accounted for by the model. If this softmax parameter is one, we are effectively assuming that the subject samples her actions from posterior beliefs about her actions; sometimes, this is called matching behavior. Alternatively, if the softmax parameter is very large, the action emitted is the action with the greatest subjective posterior—that is, the subject always chooses the most likely option. This softmax parameter can itself be estimated.

The result is the probability of the actions under the model, given a sequence of stimuli and parameters—that is, a likelihood of behavioral data given a model. Equipping the objective parameters with Gaussian priors[1] $(\theta \sim \mathcal{N}(\eta, \Pi^{(1)}))$, we can use the Laplace assumption to express a (free energy) approximation to model evidence:

$$\ln P(\tilde{u}|m, \tilde{o}) \approx \mathcal{L}(\mu) + \tfrac{1}{2}\left(\varepsilon \cdot \Pi^{(1)}\varepsilon + \ln\left|\nabla_{\mu\mu}\mathcal{L}(\mu) - \Pi^{(1)}\right|\right)$$

$$\varepsilon = \eta - \mu \qquad\qquad\qquad\qquad\qquad\qquad\qquad\qquad\qquad (9.3)$$

$$\mu = \arg\max_{\mu}\left\{\mathcal{L}(\mu) - \tfrac{1}{2}\left(\varepsilon \cdot \Pi^{(1)}\varepsilon + \ln\left|\nabla_{\mu\mu}\mathcal{L}(\mu) - \Pi^{(1)}\right|\right)\right\}$$

Equation 9.3 is the same as that unpacked in box 4.3 (generalized to a multidimensional parameter space), but here we have substituted an explicit form for the posterior covariance and assumed a normally distributed prior. In chapter 4 and in the applications in chapter 8, we ignored the terms in equation 9.3 that did not depend on the mode. However, it is important to include these here when we consider model comparison problems.

To find the value of μ that maximizes the last line of equation 9.3, we perform a gradient ascent. Under quadratic assumptions, this reduces to the following:

$$\dot{\mu} = \nabla_{\mu}\mathcal{L}(\mu) + \Pi^{(1)}\varepsilon \qquad\qquad\qquad\qquad\qquad\qquad (9.4)$$

While an explicit form for the gradient of the log likelihood used here may not be available, finite difference methods[2] may be used to calculate a reasonable numerical approximation. These may also be used to find the posterior precision, which is the second derivative (or Hessian) of the negative log likelihood plus the prior precision. Equation 9.4 is the simplest form of update, but often more sophisticated methods based on the local curvature are used.

9.4 Parametric Empirical Bayes (PEB)

The variational Laplace procedure in the previous section lets us draw inferences about, and quantify the evidence for a model of, choice behavior. This enables us to computationally phenotype an individual and to compare alternative hypotheses about that individual. However, the interesting questions often lie at a group level. For example, we might be interested in how a parameter—such as the precision of prior preferences—varies with age. To answer this question, we can use the approach of section 9.3 to fit models to the behavior of individual participants with a range of ages. We then formulate a general linear model that generates the parameter of interest, taking age into account:

$$P(\theta|\beta,X) = \mathcal{N}(X\beta, \Pi^{(2)}) \tag{9.5}$$

Here, X is a matrix whose columns are alternative explanatory variables and whose rows indicate each participant. The first column of X typically comprises a matrix of ones (to indicate the effect of the mean parameter over subjects). The second column, in our example, might be the age of each participant. The β vector indicates the size of effect of each of the explanatory variables in X. The first element of β is then the average value of the precision (or any other parameter), while the second is the effect of age on precision. This value is the slope of the line in a plot of age (x-axis) against predicted precision (y-axis). There may be an arbitrary number of columns of X, with an arbitrary number of elements in β.

Once we have fit the model expressed in equation 9.5, supplemented with priors for the β values, we can ask questions about the role of the explanatory variables. For example, we can ask whether age has an effect on the precision of prior preferences by comparing the evidence for a model in which the second element of β is allowed to deviate from zero with the evidence for a model with a precise belief that it is zero. Practically speaking, this can be done without multiple model inversions through use of Bayesian model reduction (Friston, Parr, and Zeidman 2018).

9.5   Instructions for Model-Based Analysis

In practice, we follow the steps outlined below to analyze empirical choice behavior using active inference (Schwartenbeck and Friston 2016). These refer to the relevant routines available in the SPM12 Matlab package.

1. **Collect behavioral data**, including the choices made and the sensory input available to the person making that choice. In addition, collect data of interest for second-level, between-subject analysis (e.g., whether the subject is a patient or a control subject, relevant demographic information, and so on).

2. **Formulate a POMDP model** as in chapter 7. This should be a function that takes parameters as inputs and outputs a fully specified (but not yet solved) POMDP.

3. **Specify a likelihood function** (i.e., equation 9.2). This tells us how the model should be used to calculate a likelihood. This typically calls a POMDP solver (like the `spm_MDP_VB_X.m` routine) to simulate behavior and quantifies the likelihood of observed actions.

4. **Specify prior beliefs** about the parameters in terms of expectations and precisions. Often these will be centered on zero, with precisions reflecting plausible ranges.

5. **Solve for posterior probability and model evidence.** This uses a standard inference scheme such as the variational Laplace procedure outlined above (equation 9.4). The `spm_nlsi_Newton.m` routine will do this automatically.

6. **Perform group-level analysis.** This typically makes use of PEB, which treats the estimated parameters for each individual as if they were generated by a second-level model. This allows us to test hypotheses about the causes of those parameters. Practically, this may be performed using the `spm_dcm_peb.m` routine. Alternative analyses include standard statistical tests of association between the inferred parameters for each subject and other subject-specific measures. For example, a canonical variates analysis may be used to assess the relationship between questionnaire scores and inferred parameters.

Figure 9.2's summary of these instructions are based on the behavior of the rat in the T-maze task described in chapter 7. First, we place a rat in a T-maze with a rewarding stimulus in either the left or right arm and an informative cue in the central arm. We then record the sequence of actions taken by the rat. This procedure may be repeated over multiple trials to record learned behavior, and it may be repeated for multiple different rats under different interventions (e.g., pharmacological or optogenetic).

Once these behavioral data have been obtained, we need a likelihood function that lets us quantify the probability of behavior (for a given rat in a given condition) under specific parameter settings. We can do this by formulating the POMDP model we considered in chapter 7. This must be parameterized in terms of the parameters whose likelihood we seek to find. For example, if we wanted to assess the precision associated with preferences, we might include a log scale parameter that makes the preference distribution more or less peaky.

Having set up the generative model (from the perspective of the rat), we can automatically solve the POMDP using the belief-update equations

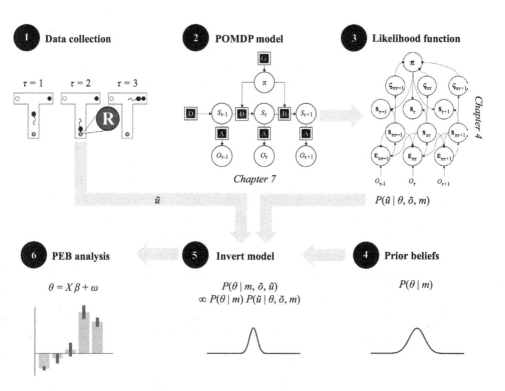

Figure 9.2
Roadmap of the six-step inversion procedure for model-based data analysis, as outlined in the main text (with reference to the chapters where more detail may be found). The arrows indicate the dependencies between each part of the process. The POMDP model must be defined for the likelihood to be evaluated. The model inversion requires collected data and the likelihood and priors; the PEB analysis cannot take place until after model inversion for each subject. Steps 4 and 5 schematize the update from a prior distribution over parameters under a model to a posterior distribution. The evidence and posterior from each model can then be combined using PEB to find posterior densities (shown as expectations with accompanying credible intervals) for the β coefficients of a linear model predicting these parameters.

in chapter 4. This lets us calculate the probability of the data (i.e., the sequence of arms visited) conditioned on the model with the (preferences) scale parameter at a particular value. Combining this likelihood with our prior completes the specification of a generative model for behavior (from the perspective of the scientist). This may be solved using variational Laplace to find a posterior probability distribution over the scale parameter for each rat.

In practice, before analyzing actual data, we may want to check the face validity of the POMDP model by using it to generate fictive data—and considering whether they are qualitatively plausible given the problem at hand. A second sensible test for the model is *parameter recovery*. This entails generating fictive data under some (known) parameterization to see whether these parameters can be recovered on inverting the model. This is useful to verify whether some parameters (or their combinations) are possible to recover (i.e., identifiable).

Finally, we can construct a design matrix for a linear model, each row of which represents a computational phenotype (e.g., a posterior density over each subject's preferences), with columns representing different attributes of those subjects. These attributes are variables that could explain differences in a rat's preferences. In addition to a column of ones indicating the average preferences over all rats, these will include things like their age, whether a drug has been administered, and so on. With this model of between-subject effects, we now perform a PEB analysis to assess the contribution of these explanatory variables to prior preferences.

9.6 · Examples of Generative Models

In this section, we leverage two examples in the literature illustrating the use of continuous and discrete generative models. First, we briefly overview the methods used by Adams, Aponte et al. (2015) and Adams, Bauer et al. (2016) (hereafter in this section, Adams et al.) to model smooth pursuit eye movements as a way of quantifying the precision parameters of each subject's generative models. An important aspect of this design was the simultaneous collection of electrophysiological data (via magnetoencephalography) that enabled the authors to ask questions about the neurobiological substrates of precision or confidence encoding. We then turn to an analysis of saccadic eye movements by Mirza et al. (2018; hereafter in this section, Mirza et al.) formulated as a POMDP model. Each of the associated experiments is cartooned in figure 9.3. Our hope is that these examples will help readers understand how the generic methods outlined above can be used empirically to answer scientific questions.

In terms of the sequence of steps outlined in figure 9.2, Adams et al. collected data (step 1) from a task in which subjects had to maintain fixation on a moving visual target. The details are not important, but this task

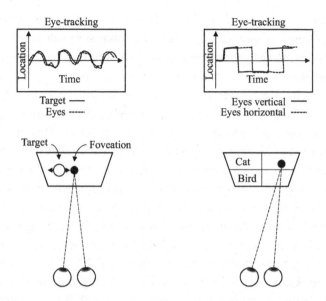

Figure 9.3
Two experiments outlined in section 9.5. The details are not important but highlight where meta-Bayesian inference has been successfully exploited and the kinds of
behavioral data it can be applied to. *Left:* Experiment of Adams et al., who measured
smooth pursuit eye movements as subjects tracked a moving target. *Right:* Experiment from Mirza et al., who measured saccadic eye movements during an exploration task. The visual display was divided into four quadrants, two of which included
stimuli (cat and bird). Different scene categories involved different configurations of
stimuli, meaning participants had to select which quadrants to foveate to gain sufficient information to categorize the scene. The Adams task (*left*) generates continuous
eye-tracking data, while the Mirza task (*right*) leads to a sequence of fixations and
may be discretized. These are the behavioral data (u) from step 1 in figure 9.2.

comprised two conditions. In the first, the target moved according to a
predictable sinusoid. In the second, it followed the same trajectory with
additive Gaussian noise. The data collected included the eye-movement
trajectories. The authors formulated a *subjective* model (step 2). Unlike the
POMDP model shown in figure 9.2, they opted for a continuous model of
the sort described in chapter 8. In brief, the model predicted proprioceptive and visual input from the eyes, where the fixation point was assumed
to be attracted to the target location. The likelihood (step 3) is constructed
using the (active) predictive coding schemes outlined in chapter 4. This
quantifies the probability of the actions (eye movements) under a set of

log precision parameters and the model set out in step 2. Moving on to step 4, the authors specified prior beliefs as normal distributions over the log precisions. They inverted the model (step 5) to find posterior distributions over these precision parameters. Step 6 did not use a PEB analysis but used the neuroimaging data collected concurrently with the behavioral task. The authors used dynamic causal modeling to estimate the gain of superficial pyramidal cells in the primary visual cortex. This means they had estimates of precision and synaptic gain for each subject. This allowed the authors to perform a group-level analysis by assessing the correlation between parameters of the subjective model and their biological substrates. Their demonstration of this correlation provides an important example of how Active Inference formulations of behavior let us ask (and answer) questions about the relationship between belief updating and neurobiology.

In our second example, Mirza et al. used the POMDP formulation of Active Inference to address the role of information gain in driving human behavior. Again, we unpack this in terms of figure 9.2's steps. Mirza et al. collected behavioral data (step 1) while subjects performed a visual foraging task. Here, the aim was to classify a visual scene into one of several groups. Each element of the scene was only revealed once subjects fixated those locations; this meant multiple fixations were required to acquire enough evidence for a given scene category. The data collected by the authors included the sequence of saccades (fast eye movements) performed. The model (step 2) used was a POMDP model described in Mirza et al. (2016) that predicted discretized proprioceptive, visual, and feedback outcomes, conditioned on the current fixation location and the scene category. Preferences (see chapter 7) were placed over the feedback outcome such that the model anticipates (and thus prefers) being correct in the categorization. The likelihood function (step 3) was obtained by solving the model using the scheme outlined in chapter 4 under different parameter settings. The parameters in question included (among others) a log scaling parameter for the precision of the preference distribution. The authors specified prior distributions (step 4) over the log scaling (and other parameters) and inverted the model for each subject (step 5). They used the log evidence estimated for each subject to assess the evidence for models that did or did not motivate behavior using the epistemic component of the expected free energy, finding greater evidence for those models that included epistemic affordance in all subjects. They then employed a PEB analysis (step 6) to assess changes in prior beliefs for subjects over the course of multiple trials, finding evidence in favor of changes in belief parameters (i.e., active learning). Finally, they used a canonical covariates analysis to assess the relationship between linear combinations of the phenotypic variables estimated for each subject (e.g., precision of preferences) and linear combinations of performance measures (e.g., percentage correct and reaction time).

9.7   Models of False Inference

Men in general are quick to believe that which they wish to be true.
—Julius Caesar

Given the relevance of these methods for fields like computational psychiatry (Friston, Stephan et al. 2014), we end with an overview of false inference, which is central to the notion of psychopathology as a failure of belief updating. One benefit of using an inferential framework like Active Inference is that it simultaneously addresses multiple dimensions of psychiatric disorders, linking together maladaptive behavior (e.g., compulsions or addictions) and psychological-level (e.g., false beliefs) and biological-level phenomena (e.g., abnormalities of neuromodulators).

As we cannot do justice here to the extensive literature that uses Active Inference in the modeling of disease processes, this section provides the briefest of overviews to suggest a framework in which to think about computational pathologies. See table 9.1 for a nonexhaustive sampling of illustrative examples, which include models specified in discrete and continuous time (based on those in chapters 7 and 8, respectively). In our discussion, we will appeal to the structure of POMDP-like models; the principles that underwrite false inference in these settings are largely the same.

The hypothesis underlying inferential approaches (like Active Inference) is that psychopathological conditions may be conceptualized as disorders of inference. The term *disorder* does not necessarily imply that the inferential mechanism is flawed (e.g., generates incorrect posterior probabilities). In most of the studies reviewed in table 9.1, the inferential mechanism operates normally, but based on a flawed generative model (i.e., a generative model endowed with aberrant prior beliefs). This means that, ultimately, pathology is a consequence of aberrant prior beliefs—and one can recover these priors using the model-based data analysis outlined in this chapter.

Table 9.1
Computational pathology

Pathology	Sources	Notes
Addiction, impulsivity, and compulsivity	FitzGerald, Schwartenbeck et al. 2015 Schwartenbeck, FitzGerald, Mathys, Dolan, Wurst et al. 2015 Mirza et al. 2019 Fradkin et al. 2020	Addiction is an important example of behavior that appears aberrant but can be framed as optimal inference under the right sort of generative model. Work by Schwartenbeck et al. illustrated this using a limited offer task wherein participants are more or less confident about whether they will receive a reward on waiting. Low confidence leads to compulsive behavior of the sort associated with addiction. Subsequent work on this theme looks at the prior beliefs associated with more or less impulsive behavior, using the patch-leaving paradigm, and examines the role of attenuated prior precision in obsessive compulsive disorder.
Delusions	Brown et al. 2013 Friston, Parr et al. 2020	Delusions, characterized by fixed, false beliefs, are simply articulated in Active Inference as precise posterior probability distributions in the absence of supportive evidence. If sufficiently precise, they will be fixed even in the face of (subsequent) contradictory evidence. The mechanisms underlying each delusion may be different. For example, failures of sensory attenuation may be central to delusions of agency. Recent work provides an example of a shared delusion (folie à deux), which depends on two agents—with no information—reaching a confident consensus about the state of the world.
Hallucinations	Adams, Stephan et al. 2013 Benrimoh et al. 2018 Parr, Benrimoh et al. 2018 Corlett et al. 2019	These simulations rest on imbalances between prior and likelihood precisions. Overinterpretation of spurious sensory data due to a failure of attenuation of likelihood precision, or a failure to correct prior beliefs due to excessive attenuation, each offer mechanisms for false perceptual inference.

Table 9.1
(continued)

Pathology	Sources	Notes
Interpersonal and personality disorders	Moutoussis et al. 2014 Prosser et al. 2018	Interpersonal inference depends on having models about other people and how they may react to our decisions. This has prompted the development of models of trust games, which rely on interactions between two (or more) parties, and charity games. The latter have been used to reproduce the self-aggrandizing and remorselessness associated with psychopathy. These traits are simulated by modulating the degree to which beliefs about self-worth depend on decisions to be charitable versus selfish and sensitivity to the approval of others.
Oculomotor syndromes	Adams, Perrinet, and Friston 2012 Parr and Friston 2018a	In these papers, continuous-time generative models are employed to predict dynamic evolution of Newtonian systems. By rendering various aspects of the generative model conditionally independent of others, oculomotor syndromes such as internuclear ophthalmoplegias may be induced.
Pharmacotherapy	Parr and Friston 2019b	Given the associations we proposed between precision parameters and neurochemicals in chapter 5, it should be possible to simulate the consequences of pharmacological manipulation of these systems. This work illustrates the consequences of several synthetic pharmacological interventions on performance of an oculomotor delay-period task, providing a proof of principle that these methods can be used to simulate not only pathology but also the influence of therapeutics.
Prefrontal syndromes	Parr, Rikhye et al. 2019	These simulations set out a difference between medial and lateral prefrontal syndromes by attenuating the precision of transitions to impair the performance of a memory guided task (lateral) versus the precision of an interoceptive likelihood that determines motivation to engage in the task (medial).

(continued)

Table 9.1
(continued)

Pathology	Sources	Notes
Visual neglect	Parr and Friston 2017a	Inattention to the left side of space may be induced by several alternative lesioned priors. Among these, an increase in Dirichlet parameters for this side of space reduces the novelty associated with saccades to the left, increasing visual sampling of the right instead. Alternatively, setting preferences consistent with right-sided proprioceptive or visual outcomes or increasing habitual engagement in right-sided saccades reproduces qualitatively similar behavior.
Disorders of interoceptive inference	Barrett et al. 2016 Allen et al. 2019 Maisto, Barca et al. 2019 Pezzulo, Maisto et al. 2019 Barca and Pezzulo 2020 Tschantz et al. 2021	Simulations of interoceptive inference (or active inference in the interoceptive domain) suggest that imbalances between prior and likelihood precisions about (for example) cardiac or gastric signals can cause false beliefs about the internal state of the body, misperceptions of bodily symptoms, and psychosomatic hallucinations. Furthermore, they can have cascading effects on autonomic regulation and action selection, causing various types of maladaptive behavior, such as hypervigilance, excessive medicine use, and excessive food restrictions.

Aberrant priors may be about states or precisions, or they may be structural priors about the form of the generative model. A useful way of thinking about the causes of pathological behavior is to think about the prior belief used for policies and about how each part of this may be disrupted to give rise to abnormal policy selection. Policy priors depend on the expected free energy, which itself depends on posterior beliefs, the potential for information gain, and prior preferences (C). Priors over policies may additionally be equipped with a fixed form term (E), representing habitual biases.

Taking each of these in turn: Posterior beliefs depend on priors and likelihoods. To form an aberrant posterior belief, one or both must be disrupted. Typically, this disruption takes the form of under- or overestimation of the balance of precisions. Excessively high likelihood, compared to prior, precision leads to overinterpretation of (potentially noisy) sensory input. This leads to *overfitting* in the sense that unwarranted conclusions may be drawn from spurious data. If the balance is disrupted in the opposite direction, favoring confidence in the prior, internally generated percepts become resistant to conflicting sensory input. Both mechanisms have been associated with the development of hallucinations, and the two can coexist when hierarchical models are employed. Given the association of various precisions with neuromodulatory chemicals (see chapter 5), it seems sensible that conditions such as Lewy body dementia, where cholinergic signaling is impaired, and schizophrenia, with abnormalities of the dopaminergic system, present with hallucinatory phenomena—that is, false perceptual inference.

Next, we consider the role of information gain. Here, the precision of the likelihood and the precision of prior beliefs tell us the degree to which uncertainty is resolvable and the amount of uncertainty there is to resolve, respectively. The precision of the prior beliefs applies to either parameters of the generative model (i.e., influences novelty) or to the states (i.e., influences salience). Interpreting the parameters of conditional probabilities as synaptic efficacies or the precisions as synaptic gains suggests that synaptic disconnection syndromes may be thought of as disruption of one or both of these. Absent synapses cannot be modulated, so this is very much like having extremely confident prior beliefs about a conditional probability, as new data cannot update the associated efficacy. This has important implications for the potential information gain of different policies, as has been exploited in modeling sensory neglect syndromes.

Finally, the preferences and policy priors provide a clear influence over behavior. These could underwrite the development of addictive habits or the apathy associated with various psychiatric and neurological syndromes (Hezemans, Wolpe, and Rowe 2020). In summary, defective prior beliefs at various places in the generative models described above provide a functional or teleological explanation for pathological behavior.

9.8 Summary

In this chapter, we outlined an approach that uses the theoretical models described in previous chapters to pose questions to empirical data. This lets us use Active Inference as a noninvasive tool to probe the computational processes that individuals use to make decisions. We have focused on a few simple examples. However, Active Inference–based models have been developed for more realistic and complicated tasks (Cullen et al. 2018) designed to evoke richer behavior for computational phenotyping. In addition to setting out a six-step process for model-based analysis, we highlighted two examples of the use of these methods. These bring out key variations in how this may proceed, including the kinds of behavior measured (smooth trajectories or discrete choices), the choice of model (continuous or discrete), and the different scientific questions being asked. The last of these is the most important, as it determines the preceding choices. We saw the use of computational phenotyping in combination with neuroimaging (Adams, Bauer et al. 2016) to ask questions about the relationship between synaptic gain and precision. In addition, we saw how model inversion may be used to assess the contributions of alternative behavioral drives and predictors of performance (Mirza et al. 2018). Ultimately, the six steps in figure 9.1 provide a generic method for designing experiments to noninvasively interrogate the implicit generative models people (or other animals) use to drive behavior. This offers an opportunity to answer questions about the function of the nervous system in health and disease.

10 Active Inference as a Unified Theory of Sentient Behavior

In general we are least aware of what our minds do best.
—Marvin Minsky

10.1 Introduction

In this chapter, we wrap up Active Inference's main theoretical points (from the first part of the book) and its practical implementations (from the second part). Then, we connect the dots: we abstract away from the specific Active Inference models discussed in previous chapters to focus on integrative aspects of the framework. One benefit of Active Inference is that it provides a complete solution to the adaptive problems that sentient organisms have to solve. It therefore offers a unified perspective on problems like perception, action selection, attention, and emotion regulation, which are usually treated in isolation in psychology and neuroscience—and addressed using distinct computational approaches in artificial intelligence. We will discuss the Active Inference perspective on each of these problems (and more) in the context of established theories, such as cybernetics, ideomotor theory of action, reinforcement learning, and optimal control. Finally, we briefly discuss how the scope of Active Inference can be extended to cover other biological, social, and technological topics that are not discussed in depth in this book.

10.2 Wrapping Up

This book offers a systematic account of the theoretical underpinnings and practical implementations of Active Inference. Here, we briefly summarize the discussion of the first nine chapters. This offers an opportunity to rehearse the key constructs of Active Inference that will be useful in the remainder of this chapter.

In chapter 1, we introduced Active Inference as a normative approach to understanding sentient creatures that form part of action-perception loops with their environment (Fuster 2004). We explained that normative approaches start from first principles to derive and test empirical predictions about the phenomenon of interest—here, the ways living organisms persist while engaging in adaptive exchanges (*action-perception loops*) with their environment. We also considered that one could arrive at Active Inference by following a low road or a high road.

In chapter 2, we illustrated the low road to Active Inference. This road starts from the idea that the brain is a prediction machine, endowed with a *generative model*: a probabilistic representation of how hidden causes in the world generate sensations (e.g., how light reflected off an apple stimulates the retina). By inverting this model, it infers the causes of its sensations (e.g., whether I am seeing an apple, given that my retina is stimulated in a certain way). This view of perception (aka perception-as-inference) has its historical roots in the Helmholtzian notion of unconscious inference and, more recently, in the *Bayesian brain* hypothesis. Active Inference extends this view by bringing action control and planning within the compass of inference (aka control-as-inference, planning-as-inference). Most importantly, it shows that perception and action are not quintessentially separable processes but fulfill the same objective. We first described this objective more informally, as the minimization of a discrepancy between one's model and the world (which generally reduces to surprise or prediction error minimization). Put simply, one can minimize the discrepancy between a model and the world in two ways: by changing one's mind to fit the world (perception) or by changing the world to fit the model (action). These can be described in terms of Bayesian inference. However, exact inference is often intractable, so Active Inference uses a (variational) approximation (noticing that exact inference may be seen as a special case of approximate inference). This leads to the second, more formal description of the common objective of perception and action, as *variational free energy* minimization. This is the core quantity used in Active Inference and may be unpacked in terms of its constituent parts (e.g., energy and entropy, complexity and accuracy, or surprise and divergence). Finally, we introduced a second kind of free energy: *expected free energy*. This is particularly important during planning, as it affords a way to score alternative policies by considering the future outcome that they are expected to generate. This too may be unpacked in terms of its constituent parts (e.g., information gain and pragmatic value, expected ambiguity and risk).

In chapter 3, we illustrated the high road to Active Inference. This alternative road starts from the deflationary imperative for biological organisms to preserve their integrity and avoid dissipation, which can be described as avoiding surprising states. We then introduced the notion of a *Markov blanket*: a formalization of the statistical separation between the organism's internal states and the world's external states. Crucially, internal and external states can only influence each other vicariously via intermediate (active and sensory) variables, called blanket states. This statistical separation—mediated by the Markov blanket—is crucial to endowing an organism with some degree of autonomy from the external world. To understand why this is a useful perspective, consider the following three consequences.

First, an organism with a Markov blanket appears to model the external environment in a Bayesian sense: its internal states correspond—on average—to an approximate posterior belief about external states of the world. Second, the autonomy is guaranteed by the fact that the organism's model (its internal states) is not unbiased but prescribes some existential preconditions (or prior preferences) that must be maintained—for example, for a fish, being in the water. Third, equipped with this formalism, it is possible to describe optimal behavior (with respect to prior preferences) as the maximization of (Bayesian) *model evidence* by perception and action. By maximizing model evidence (i.e., *self-evidencing*) an organism ensures that it realizes its prior preferences (e.g., a fish stays in the water) and avoids surprising states. In turn, the maximization of model evidence is (approximately) mathematically equivalent to the minimization of *variational free energy*—hence we arrive again (in another way) at the same central construct of Active Inference discussed in chapter 2. Finally, we detailed the relationship between minimizing surprise and *Hamilton's principle of least Action*. This evinces the formal relationship between Active Inference and first principles in statistical physics.

In chapter 4, we outlined the formal aspects of Active Inference. We focused on the passage from Bayesian inference to a tractable approximation—*variational inference*—and the resulting objective for organisms to minimize *variational free energy* via perception and action. The insight from this treatment is the importance of the generative model that creatures use to make sense of their world. We introduced two kinds of generative models that express our beliefs about how data are generated, using discrete or continuous variables. We explained that both afford the same Active Inference, but they apply when states of affairs are formulated in discrete time (as partially observed Markov decision problems) or continuous time (as stochastic differential equations), respectively.

In chapter 5, we remarked on the difference between the normative principle of free energy minimization and a process theory about how this principle may be implemented by the brain—and explained that the latter generates testable predictions. We then outlined aspects of the process theories accompanying Active Inference, which encompass domains such as neuronal message passing, including neuroanatomical circuitry (e.g., cortico-subcortical loops) and neuromodulation. For example, at an anatomical level, message passing maps nicely to a canonical cortical microcircuit, with predictions that stem from deep cortical layers at one level and target superficial cortical layers at the level below (Bastos et al. 2012). At a more systemic level, we discussed how Bayesian inference, learning, and precision weighting correspond to neuronal dynamics, synaptic plasticity, and neuromodulation, respectively, and how the top-down and bottom-up neural message passing of predictive coding maps to slower (e.g., alpha or beta) and faster (e.g., gamma) brain rhythms. These and other examples illustrate that after designing a specific Active Inference model, one can draw neurobiological implications from the form of its generative model.

In chapter 6, we provided a recipe to design Active Inference models. We saw that while all creatures minimize their variational free energy, they behave in different, sometimes opposite ways because they are endowed with different generative models. Therefore, what distinguishes different (e.g., simpler from more complex) creatures is just their generative model. There is a rich repertoire of possible generative models, which correspond to different biological (e.g., neuronal) implementations and produce different adaptive—or maladaptive—behaviors in different contexts and ecological niches. This renders Active Inference equally appropriate for characterizing simple creatures like bacteria that sense and seek nutrient gradients, complex creatures like us that pursue sophisticated goals and engage in rich cultural practices, or even different individuals—to the extent that ones appropriately characterizes their respective generative models. Evolution appears to have discovered increasingly sophisticated design structures for brains and bodies that made organisms able to deal with (and shape) rich ecological niches. Modelers can reverse-engineer this process and specify the designs for brains and bodies of creatures of interest, in terms of generative models, based on the kinds of niche they occupy. This corresponds to a series of design choices (e.g., models using discrete or categorical variables, shallow or hierarchical models)—which we unpacked in the chapter.

In chapters 7 and 8, we provided numerous examples of Active Inference models in discrete and continuous time, which address problems of perceptual inference, goal-directed navigation, model learning, action control, and more. These examples were designed to showcase the variety of emergent behaviors under these models and to detail the principles of how they are specified practically.

In chapter 9, we discussed how to use Active Inference for model-based data analysis and to recover the parameters of an individual's generative model, which better explain the subject's behavior in a task. This *computational phenotyping* uses the same form of Bayesian inference discussed in the rest of the book, but in a different way: it helps design and evaluate (objective) models of others' (subjective) models.

10.3 Connecting the Dots: The Integrative Perspective of Active Inference

Some decades ago, the philosopher Dennett lamented that cognitive scientists devoted too much effort to modeling isolated subsystems (e.g., perception, language understanding) whose boundaries are often arbitrary. He suggested to try instead modeling "the whole iguana": a complete cognitive creature (perhaps a simple one) and an environmental niche for it to cope with (Dennett 1978).

One benefit of Active Inference is that it offers a first principle account of the ways in which organisms solve their adaptive problems. The normative approach pursued in this book assumes that it is possible to start from the principle of variational free energy minimization and derive implications about specific cognitive processes, such as perception, action selection, attention and emotion regulation, and their neuronal underpinnings.

Imagine a simple creature that must solve problems like finding food or shelter. When cast as Active Inference, the creature's problems can be described in enactive terms, as acting to solicit preferred sensations (e.g., food-related sensations). To the extent that these preferred sensations are included (as prior beliefs) in its generative model, the organism is effectively gathering evidence for its model—or, more allegorically, for its existence (i.e., maximizing model evidence or self-evidencing). This simple principle has ramifications for psychological functions traditionally considered in isolation, such as perception, action control, memory, attention, intention, emotion, and more. For example, perception and action are both self-evidencing, in the sense that a creature can align what it expects, given its generative model, with what it senses either by changing its beliefs (about the presence of food) or by changing the world (soliciting food-related sensations). Memory and attention can also be thought of as optimizing the same objective. Long-term memory develops through learning the parameters of a generative model. Working memory is belief updating when beliefs are about external states in the past and future. Attention is the optimization of beliefs about the precision of sensory input. Forms of planning (and intentionality) can be conceptualized by appealing to the capacity of (some) creatures to select among alternative futures, which in turn requires temporally deep generative models. These predict the outcomes that would result from a course of action and are optimistic about these outcomes. This optimism manifests as the belief that future outcomes will lead to preferred outcomes. Deep temporal models can also help us understand sophisticated forms of prospection (where beliefs about the present are used to derive beliefs about the future) and retrospection (where beliefs about the present are used to update beliefs about the past). Forms of interoceptive regulation and emotion can be conceptualized by appealing to generative models of internal physiology that predict the allostatic consequences of future events.

As the above examples illustrate, there is an important consequence of studying cognition and behavior from the perspective of a normative theory of sentient behavior. Such theory does not start by assembling separate cognitive functions, such as perception, decision-making, and planning. Rather, it starts by providing a complete solution to the problems that organisms have to solve and then analyzing the solution to derive implications about cognitive functions. For example, which mechanisms permit a living organism or artificial creature (e.g., a robot) to perceive the world, remember it, or plan (Verschure et al. 2003, 2014; Verschure 2012; Pezzulo, Barsalou et al. 2013; Krakauer et al. 2017)? This is an important move as the taxonomies of cognitive functions—used in psychology and neuroscience textbooks—largely inherit from early philosophical and psychological theories (sometimes called *Jamesian categories*). Despite their great heuristic value, they may be quite arbitrary—or they may not correspond to separate cognitive and neural processes (Pezzulo and Cisek 2016, Buzsaki 2019, Cisek 2019). Indeed, these Jamesian categories may be candidates for how our generative models explain our engagement with the sensorium—as opposed to explaining that engagement. For example, the solipsistic hypothesis that "I am perceiving" is just my explanation for current states of affairs that include my belief updating.

Adopting a normative perspective may also help in identifying formal analogies between cognitive phenomena studied in different domains. One example is the trade-off between exploration and exploitation, which appears in various guises (Hills et al. 2015). This trade-off is often studied during foraging, when creatures must choose between exploiting previous successful plans and exploring novel (potentially better) ones. However, the same trade-off occurs during memory search and deliberation with limited resources (e.g., time limitations or search effort), when creatures have the choice between exploiting their current best plan versus investing more time and cognitive effort to explore additional possibilities. Characterizing these apparently disconnected phenomena in terms of free energy can potentially reveal deep similarities (Friston, Rigoli et al. 2015; Pezzulo, Cartoni et al. 2016; Gottwald and Braun 2020).

Finally, in addition to a unified perspective on psychological phenomena, Active Inference offers a principled means of understanding the corresponding neural computations. In other words, it offers a process theory that connects cognitive processing to (expected) neuronal dynamics. Active Inference assumes that everything that matters about brains, minds, and behavior can be described in terms of the minimization of variational free energy. In turn, this minimization has specific neural signatures (in terms of, e.g., message passing or brain anatomy) that can be empirically validated.

In the rest of this chapter, we explore some implications of Active Inference for psychological functions—as if we were sketching a psychology textbook. For each of these functions, we also highlight some points of contact (or divergence) between Active Inference and other popular theories in the literature.

10.4   Predictive Brains, Predictive Minds, and Predictive Processing

I have this picture of pure joy
it's of a child with a gun
he's aiming straight in front of himself,
shooting at something that isn't there.
—Afterhours, "Quello che non c'è" (Something that isn't there)

Traditional theories of brain and cognition emphasize feedforward transductions from external stimuli to internal representations and then motor actions. This has been called a "sandwich model," as everything that is in between stimuli and responses is assigned the label "cognitive" (Hurley 2008). In this perspective, the main function of the brain is to transform incoming stimuli into contextually appropriate responses.

Active Inference departs significantly from this view by emphasizing predictive and goal-directed aspects of brain and cognition. In psychological terms, Active Inference creatures (or their brains) are *probabilistic inference machines*, which continuously generate predictions based on their generative models.

Self-evidencing creatures use their predictions in two fundamental ways. First, they compare predictions with incoming data to validate their hypotheses (predictive coding) and—at a slower timescale—revise their models (learning). Second, they enact predictions to guide the ways they gather data (Active Inference). By doing so, Active Inference creatures fulfill two imperatives: epistemic (e.g., visually exploring places where salient information is present that can resolve uncertainty about hypotheses or models) and pragmatic (e.g., moving to locations where preferred observations such as rewards can be secured). The epistemic imperative renders both perception and learning *active* processes, whereas the pragmatic imperative renders behavior *goal directed*.

10.4.1   Predictive Processing
This predictive- and goal-centric view of brain—and cognition—is closely related to (and provided inspiration for) *predictive processing* (PP): an emerging framework in philosophy of mind and epistemology, which sees prediction as central to brain and cognition and appeals to concepts of "predictive brains" or "predictive minds" (Clark 2013, 2015; Hohwy 2013).

Sometimes PP theories appeal to the specific functioning of Active Inference and some of its constructs, such as generative models, predictive coding, free energy, precision control, and Markov blankets, but they sometimes appeal to other constructs, such as coupled inverse and forward models, which are not part of Active Inference. Therefore, the term *predictive processing* is used in a broader (and less constrained) sense compared to Active Inference.

Predictive processing theories have attracted considerable attention in philosophy, given their potential for unification in many senses: across multiple domains of cognition, including perception, action, learning, and psychopathology; from lower (e.g., sensorimotor) to higher levels of cognitive processing (e.g., psychological constructs); from simple biological organisms to brains, individuals, and social and cultural constructs. Another appeal of PP theories is that they make use of conceptual terms, such as *beliefs* and *surprise*, which speak to a psychological level of analysis familiar to philosophers (with the caveat that sometimes these terms may have technical meanings that differ from common usage).

Yet, as the interest in PP grows, it has become apparent that philosophers have different opinions on its theoretical and epistemological implications. For example, it has been interpreted in internalist (Hohwy 2013), embodied or action-based (Clark 2015), and enactivist and nonrepresentational terms (Bruineberg et al. 2016, Ramstead et al. 2019). The debate around these conceptual interpretations goes beyond the scope of this book.

10.5 Perception

You can't depend on your eyes when your imagination is out of focus.
—Mark Twain

Active Inference considers perception as an inferential process based on a generative model of how sensory observations are generated. Bayes' rule essentially inverts the model to compute a belief about the hidden state of the environment, given the observations. This idea of perception-as-inference dates back to Helmholtz (1866) and was often reproposed in psychology, computational neuroscience, and machine learning (e.g., analysis-by-synthesis) (Gregory 1980, Dayan et al. 1995, Mesulam 1998, Yuille and Kersten 2006). This generative modeling approach has been demonstrated to be effective in facing challenging perceptual problems, such as breaking text-based CAPT-CHAs (George et al. 2017).

10.5.1 Bayesian Brain Hypothesis

The most prominent contemporary expression of this idea is the Bayesian brain hypothesis, which has been applied to several domains such as decision-making, sensory processing, and learning (Doya 2007). Active Inference provides a normative foundation to these inferential ideas by deriving them from the imperative of minimizing variational free energy. As the same imperative extends to action dynamics, Active Inference naturally models *active perception* and the ways in which organisms actively sample observations to test their hypotheses (Gregory 1980). Under the Bayesian brain agenda, instead, perception and action are modeled in terms of different imperatives (where action requires Bayesian decision theory; see section 10.7.1).

More broadly, the Bayesian brain hypothesis refers to a family of approaches that are not necessarily integrated and often make different empirical predictions. These include, for example, the computational-level proposal that the brain performs Bayes-optimal sensorimotor and multisensory integration (Kording and Wolpert 2006), the algorithmic-level proposal that the brain implements specific approximations of Bayesian inference, such as decision-by-sampling (Stewart et al. 2006), and the neural-level proposals about the specific ways in which neural populations may perform probabilistic computations or encode probability distributions—for example, as samples or probabilistic population codes (Fiser et al. 2010, Pouget et al. 2013). At each level of explanation, there are competing theories on the field. For example, it is common to appeal to approximations of exact Bayesian inference to explain deviations from optimal behavior, but different works consider different (and not always compatible) approximations, such as different sampling approaches. More broadly, the relations between proposals at different levels are not always straightforward. This is because Bayesian computations can be realized (or approximated) in multiple algorithmic ways, even without explicitly representing probability distributions (Aitchison and Lengyel 2017).

Active Inference provides a more integrated perspective that connects normative principles and process theories. At the normative level, its central assumption is that all processes minimize variational free energy. The corresponding process theory for inference uses a gradient descent on free energy, which has clear neurophysiological implications, explored in chapter 5 (Friston, FitzGerald et al. 2016). More broadly, one can start from the principle of free energy minimization to derive implications about brain architectures.

For example, the canonical process model of perceptual inference (in continuous time) is *predictive coding*. Predictive coding was initially proposed as a theory of hierarchical perceptual processing by Rao and Ballard (1999) to explain a range of documented top-down effects, which were difficult to reconcile with feedforward architectures as well as known physiological facts (e.g., the existence of forward, or bottom-up, and backward, or top-down, connections in sensory hierarchies). However, predictive coding can be derived from the principle of free energy minimization, under some assumptions, such as the Laplace approximation (Friston 2005). Furthermore, Active Inference *in continuous time* can be constructed as a directed extension of predictive coding into the domain of action—by endowing a predictive coding agent with motor reflexes (Shipp et al. 2013). This leads us to the next point.

10.6 Action Control

> If you can't fly then run, if you can't run then walk, if you can't walk then crawl, but whatever you do you have to keep moving forward.
> —Martin Luther King

In Active Inference, action processing is analogous to perceptual processing, as both are guided by forward predictions—exteroceptive and proprioceptive, respectively. It is the (proprioceptive) prediction that "my hand grasps the cup" that induces a grasping movement. The equivalence between action and perception exists also at the neurobiological level: the architecture of the motor cortex is organized in the same way as the sensory cortex—as a predictive coding architecture, with the exceptions that it can influence motor reflexes in the brain stem and spine (Shipp et al. 2013) and that it receives relatively little ascending input. Motor reflexes permit controlling movement by setting "equilibrium points" along a desired trajectory—an idea that corresponds to the *equilibrium point hypothesis* (Feldman 2009).

Importantly, initiating an action—like grasping a cup—requires regulation of the precision (inverse variance) of prior beliefs and sensory streams appropriately. This is because the relative values of these precisions determine the way in which a creature manages the conflict between its prior belief (that it holds the cup) and its sensory input (signaling that it does not). An imprecise prior belief about grasping a cup can be easily revised in the light of conflicting sensory evidence—producing a change of mind and no action. Rather, when the prior belief dominates (i.e., has higher precision), it is maintained even in the face of conflicting sensory evidence—and it induces a grasping action to resolve the conflict. To ensure that this is the case, action initiation induces a transient sensory attenuation (or downweighting sensory prediction errors). Failure of this sensory attenuation can have maladaptive consequences, such as the failure to initiate or control movements (Brown et al. 2013).

10.6.1 Ideomotor Theory
In Active Inference, action stems from (proprioceptive) predictions and not motor commands (Adams, Shipp, and Friston 2013). This idea connects Active Inference to *ideomotor theory* of action: a framework to understand action control that dates back to William James (1890) and the later theories of "event coding" and "anticipatory behavioural control" (Hommel et al. 2001, Hoffmann 2003). Ideomotor theory suggests that *action-effect links* (similar to forward models) are key mechanisms in the architecture of cognition. Importantly, these links can be used bidirectionally. When they are used in the *action-effect* direction, they permit generating sensory predictions; when they are used in the *effect-action* direction, they permit selecting actions that achieve desired perceptual consequences—implying that actions are selected and controlled on the basis of their predicted consequences (hence the term *ideo + motor*). This anticipatory view of action control is supported by a body of literature that documents the effects of (anticipated) action consequences on action selection and execution (Kunde et al. 2004). Active Inference provides a mathematical characterization of this idea that also includes additional mechanisms, such as the importance of precision control and sensory attenuation, which are not fully investigated in (but are compatible with) ideomotor theory.

10.6.2 Cybernetics

Active Inference is closely related to cybernetic ideas about the purposeful, goal-directed nature of behavior and the importance of (feedback-based) agent-environment interactions, as exemplified by the TOTE (Test, Operate, Test, Exit) and related models (Miller et al. 1960; Pezzulo, Baldassarre et al. 2006). In both TOTE and Active Inference, the selection of actions is determined by the discrepancy between a preferred (goal) state and the current state. These approaches diverge from simple stimulus-response relationships, as more commonly assumed in behaviorist theory and computational frameworks like reinforcement learning (Sutton and Barto 1998).

The notion of action control in Active Inference is particularly akin to *perceptual control theory* (Powers 1973). Central to perceptual control theory was the notion that what is controlled is a perceptual state, not a motor output or action. For example, while driving, what we control—and keep stable over time in the face of disturbances—is our reference or desired velocity (e.g., 90 mph), as signaled by the speedometer, whereas the actions we select for this (e.g., accelerating or decelerating) are more variable and context dependent. For example, depending on the disturbance (e.g., wind, a steep road, or other cars), we would need to either accelerate or decelerate to maintain the reference velocity. This view implements William James's (1890) suggestion that "humans achieve stable goals via flexible means."

While in both Active Inference and perceptual control theory it is a perceptual (and specifically a proprioceptive) prediction that controls action, the two theories differ in how control is operated. In Active Inference but not perceptual control theory, action control has anticipatory or feedforward aspects, based on generative models. In contrast, perceptual control theory assumes that feedback mechanisms are largely sufficient to control behavior, whereas trying to predict a disturbance, or exerting feedforward (or open-loop) control, is worthless. However, this objection was mainly intended to address the limitations of control theories that use inverse-forward models (see next section). Under Active Inference, generative or forward models are not used to predict a disturbance but to predict future (desired) states and trajectories to be fulfilled by acting—and to infer the latent cause of perceptual events.

Finally, another important point of contact between Active Inference and perceptual control theory is the way they conceptualize control hierarchies. Perceptual control theory proposes that higher hierarchical levels control lower hierarchical levels by setting their reference points or set-points (i.e., *what* they have to achieve) by leaving them free to select the means to achieve them rather than by setting or biasing the actions that the lower levels have to perform (i.e., *how* to operate). This stands in contrast with most theories of hierarchical and top-down control, in which higher levels either directly select plans (Botvinick 2008) or bias the selection of actions or motor commands at lower hierarchical levels (Miller and Cohen 2001). Similar to perceptual control theory, in Active Inference one can decompose hierarchical control in terms of a (top-down) cascade of goals and subgoals, which can be autonomously achieved at the appropriate (lower) levels. Furthermore, in Active Inference, the contribution of goals represented at different levels of the control hierarchy can be modulated (precision weighted) by motivational processes, in such a way that the more salient or urgent goals are prioritized (Pezzulo, Rigoli, and Friston 2015, 2018).

10.6.3   Optimal Control Theory

The way Active Inference accounts for action control is significantly different from other models of control in neuroscience, such as *optimal control theory* (Todorov 2004, Shadmehr et al. 2010). This framework assumes that the brain's motor cortex selects actions using a (reactive) control policy that maps stimuli to responses. Active Inference, instead, assumes that the motor cortex conveys predictions, not commands.

Furthermore, while both optimal control theory and Active Inference appeal to internal models, they describe internal modeling in different ways (Friston 2011). In optimal control, there is a distinction between two kinds of internal models: *inverse models* encode stimulus-response contingencies and select motor commands (according to some cost function), whereas *forward models* encode action-outcome contingencies and provide inverse models with simulated inputs to replace noisy or delayed feedback, hence going beyond a pure feedback control scheme. Inverse and forward models can also operate in a loop that is detached from external action-perception (i.e., when inputs and outputs are suppressed) to support internal, "what if" simulations of action sequences. Such internal simulations of action have been linked to various cognitive functions, such as planning, action perception, and imitation in social domains (Jeannerod 2001, Wolpert et al. 2003) as well as various disorders of movement and psychopathologies (Frith et al. 2000).

In contrast to the forward-inverse modeling scheme, in Active Inference forward (generative) models do the heavy lifting of action control, whereas inverse models are minimalistic and often reduce to simple reflexes resolved at the peripheral level (i.e., in the brain stem or spinal cord). Action is initiated when there is a difference between anticipated and observed states (e.g., desired, current arm positions)—that is, a sensory prediction error. This means a motor command is equivalent to a prediction made by the forward model as opposed to something computed by an inverse model as in optimal control. The sensory (more precisely, proprioceptive) prediction error is resolved by an action (i.e., arm movement). The gap to be filled by action is considered so small that it does not require a sophisticated inverse model but a much simpler motor reflex (Adams, Shipp, and Friston 2013).[1] What renders a motor reflex simpler than an inverse model is that it does not encode a mapping from inferred states of the world to action but a much simpler mapping between action and sensory consequences. See Friston, Daunizeau et al. (2010) for further discussion.

Another crucial difference between optimal motor control and Active Inference is that the former uses a notion of *cost* or *value function* to motivate action, whereas the latter replaces it with the Bayesian notion of prior (or prior preference, implicit in expected free energy)—as we discuss in the next section.

10.7   Utility and Decision-Making

Action expresses priorities.
—Mahatma Gandhi

The notion of a cost or value function of states is central in many fields, such as optimal motor control, economic theories of utility maximization, and reinforcement learning. For example, in optimal control theory, the optimal control policy for a reaching task is often defined as the one that minimizes a specific cost function (e.g., is smoother or has minimum jerk). In reinforcement learning problems, such as navigating in a maze that includes one or more rewards, the optimal policy is the one that permits maximizing (discounted) reward while also minimizing movement costs. These problems are often solved using the Bellman equation (or the Hamilton-Jacobi-Bellman equation in continuous time), whose general idea is that the value of a decision can be decomposed in two parts: the immediate reward and the value of the remaining part of the decision problem. This decomposition affords the iterative procedure of *dynamic programming*, which is at the core of control theory and reinforcement learning (RL) (Bellman 1954).

Active Inference differs from the above approach in two main ways. First, Active Inference does not consider utility maximization alone but the broader objective of (expected) free energy minimization, which also includes additional (epistemic) imperatives, such as the disambiguation of current state and novelty seeking (see figure 2.5). These additional objectives are sometimes added on to classical rewards—for example, as a "novelty bonus" (Kakade and Dayan 2002) or "intrinsic reward" (Schmidhuber 1991, Oudeyer et al. 2007, Baldassarre and Mirolli 2013, Gottlieb et al. 2013)—but they arise automatically in Active Inference, enabling it to resolve exploration-exploitation trade-offs implicit in many decisions. The reason for this is that free energies are functionals of beliefs, which means we are in the realm of belief optimization as opposed to external reward functions. This is essential in explorative problems, wherein success depends on resolving as much uncertainty as possible.

Second, in Active Inference, the notion of cost is absorbed into the prior. The prior (or prior preference) specifies an objective for control—for example, a trajectory to follow or an endpoint to reach. Using priors to encode preferred observations (or sequences) may be more expressive than using utilities (Friston, Daunizeau, and Kiebel 2009). Using this method, finding the optimal policy is recast as a problem of inference (of a sequence of control states that realize the preferred trajectory) and does not require a value function or the Bellman equation—although can appeal to a similar recursive logic (Friston, Da Costa et al. 2020). There are at least two fundamental differences between the ways priors and value functions are normally used in Active Inference and RL, respectively. First, RL methods use value functions of states or of state-action pairs—whereas Active Inference uses priors over observations. Second, value functions are defined in terms of the expected return of being in a state (or performing an action in a state) following a specific policy—that is, the sum of future (discounted) rewards obtained by starting in the state and then executing the policy. In contrast, in Active Inference, priors do not usually sum future rewards, nor do they discount them. Rather, something analogous to the expected return only emerges in Active Inference when the expected free energy of a policy is calculated. The implication is that expected free energy is the closest analogue to the value function. However, even this differs in the sense that expected free energy is a functional of beliefs about states, not a function of states. Having said this, it is possible to construct priors that resemble value functions of states in RL—for example, by caching expected free energy calculations in these states (Friston, FitzGerald et al. 2016; Maisto, Friston, and Pezzulo 2019).

Furthermore, absorbing the notion of utility into the prior has an important theoretical consequence: priors play the role of goals and render the generative model biased—or optimistic, in the sense that the creature believes it will encounter preferred outcomes. It is this optimism that underwrites inferred plans that achieve desired outcomes in Active Inference; a failure of this sort of optimism may correspond to apathy (Hezemans et al. 2020). This stands in contrast with other formal approaches to decision-making, such as Bayesian decision theory, which separate the probability of events from their utility. Having said this, this distinction is somewhat superficial, as a utility function can always be rewritten as encoding a prior belief, consistent with the fact that behaviors that maximize a utility function are a priori (and by design) more probable. From one (slightly tautological) deflationary perspective, this is the definition of utility.

10.7.1 Bayesian Decision Theory

Bayesian decision theory is a mathematical framework that extends the ideas of the Bayesian brain (discussed above) to the domains of decision-making, sensorimotor control, and learning (Kording and Wolpert 2006, Shadmehr et al. 2010, Wolpert and Landy 2012). Bayesian decision theory describes decision-making in terms of two distinct processes. The first process uses Bayesian computations to predict the probability of future (action- or policy-dependent) outcomes, and the second process defines the preference over plans, using a (fixed or learned) utility or cost function. The final decision (or action selection) process integrates both streams, thus selecting (with higher probability) the action plan that has the higher probability of yielding the higher reward. This stands in contrast to Active Inference, in which the prior distribution directly signals what is valuable for the organism (or what has been valuable during evolutionary history). However, parallels could be drawn between the two streams of Bayesian decision theory and the optimization of variational and expected free energy, respectively. Under

Active Inference, the minimization of variational free energy affords accurate (and simple) beliefs about the state of the world and its likely evolution. The prior belief that expected free energy will be minimized through policy selection incorporates the notion of preferences.

In some circles, there are concerns about the status of Bayesian decision theory. This follows from the complete class theorems (Wald 1947, Brown 1981) that say for any given pair of decisions and cost functions, there exist some prior beliefs that render the decisions Bayes optimal. This means that there is an implicit duality or degeneracy when dealing separately with prior beliefs and cost functions. In one sense, Active Inference resolves this degeneracy by absorbing utility or cost functions into prior beliefs in the form of preferences.

10.7.2 Reinforcement Learning

Reinforcement learning (RL) is an approach to solving Markov decision problems that is popular in both artificial intelligence and the cognitive sciences (Sutton and Barto 1998). It focuses on how agents learn a policy (e.g., pole balancing strategy) by trial and error: by trying out actions (e.g., move to the left) and receiving positive or negative reinforcements, depending on action success (e.g., pole balanced) or failure (e.g., pole fallen).

Active Inference and RL address overlapping sets of problems but differ in many respects mathematically and conceptually. As noted above, Active Inference dispenses with the notions of reward, value functions, and Bellman optimality that are key to reinforcement learning approaches. Furthermore, the notion of *policy* is used differently in the two frameworks. In RL a policy denotes a set of stimulus-response mappings that need to be learned. In Active Inference, a policy is part of the generative model: it denotes a sequence of control states that need to be inferred.

Reinforcement learning approaches are plentiful, but they can be subdivided into three main families. The first two methods try to learn good (state or state-action) value functions, albeit in two different ways.

Model-free methods of RL learn value functions directly from experience: they perform actions, collect rewards, update their value functions, and use them to update their policies. The reason they are called *model-free* is because they do not use a (transition) model that permits predicting future states—of the sort used in Active Inference. Instead, they implicitly appeal to simpler kinds of models (e.g., state-action mappings). Learning value functions in model-free RL often involves computing *reward* prediction errors, as in the popular temporal-difference rule. While Active Inference often appeals to prediction errors, these are *state* prediction errors (as there is no notion of reward in Active Inference).

Model-based methods of RL do not learn value functions or policies directly from experience. Rather, they learn a model of the task from experience, use the model to plan (simulate possible experiences), and update value functions and policies from these simulated experiences. While both Active Inference and reinforcement learning appeal to model-based planning, they use it differently. In Active Inference, planning is the computation of the expected free energy for each policy, not a means to update value functions. Arguably, if the expected free energy is seen as a value functional, it could be said that inferences drawn using the generative model are used to update this functional—offering a point of analogy between these approaches.

The third family of RL approaches, *policy gradient methods*, tries to optimize policies directly, without intermediate value functions, which are central to both model-based and model-free RL. These methods start from parameterized policies, able to generate (for example) movement trajectories, and then optimizes them by changing the parameters to increase (decrease) the likelihood of a policy if the trajectory results in a high (low) positive reward. This approach relates policy gradient methods to Active Inference, which also dispenses with value functions (Millidge 2019). However, the general objective of policy gradients (maximizing long-term cumulative reward) differs from Active Inference.

Besides the formal differences between Active Inference and RL, there are also several important conceptual differences. One difference regards how the two approaches interpret goal-directed and habitual behavior. In the animal learning literature, goal-directed choices are mediated by the (prospective) knowledge of the contingency between an action and its outcome (Dickinson and Balleine 1990), whereas habitual choices are not prospective and depend on simpler (e.g., stimulus-response) mechanisms. A popular idea in RL is that goal-directed and habitual choices correspond to *model-based* and *model-free* RL, respectively, and that these are acquired in parallel and continuously compete to control behavior (Daw et al. 2005).

Active Inference instead maps goal-directed and habitual choices to different mechanisms. In Active Inference (in discrete time), policy selection is quintessentially model-based and hence fits the definition of goal-directed, deliberative choices. This is similar to what happens in model-based RL, but with a difference. In model-based RL, actions are *selected* in a prospective manner (using a model) but are *controlled* in a reactive way (using stimulus-response policies); in Active Inference, actions can be controlled in a proactive way—through fulfilling proprioceptive predictions (on action control, see section 10.6).

In Active Inference, habits can be acquired by executing goal-directed policies and then caching information about which policies are successful in which contexts. The cached information can be incorporated as a prior value of policies (Friston, FitzGerald et al. 2016; Maisto, Friston, and Pezzulo 2019). This mechanism permits executing policies that have a high prior value (in a given context) without deliberation. This can be thought of simply as observing "what I do" and learning that "I am the sort of creature that tends to do this" over multiple exposures to a task. In contrast to model-free RL, where habits are acquired independently of goal-directed policy selection, in Active Inference habits are acquired by repeatedly pursuing goal-directed policies (e.g., by caching their results).

In Active Inference, goal-directed and habitual mechanisms can cooperate rather than only compete. This is because the prior belief over policies depends on both a habitual term (a prior value of policies) and a deliberative term (expected free energy). Hierarchical elaborations of Active Inference suggest that reactive and goal-directed mechanisms could be arranged in a hierarchy rather than as parallel pathways (Pezzulo, Rigoli, and Friston 2015).

Finally, it is worth noting that Active Inference and RL differ subtly in how they conceive behavior and its causes. RL originates from behaviorist theory and the idea that behavior results from trial-and-error learning mediated by reinforcement. Active Inference assumes instead that behavior is the result of an inference. This leads us to the next point.

10.7.3 Planning as Inference

In the same way that it is possible to cast perceptual problems as problems of inference, it is also possible to cast control problems in terms of (approximate) Bayesian inference (Todorov 2008). In keeping with this, in Active Inference, planning is seen as an inferential process: the inference of a sequence of control states of the generative model.

This idea is closely related to other approaches, which include *control-as-inference* (Rawlik et al. 2013, Levine 2018), *planning-as-inference* (Attias 2003, Botvinick and Toussaint 2012), and *risk-sensitive* and *KL control* (Kappen et al. 2012). In these approaches, planning proceeds through inferring a posterior distribution over actions, or sequences of actions, using a dynamic generative model that encodes probabilistic contingencies between states, actions, and future (expected) states. The best action or plan can be inferred by conditioning the generative model on observing future rewards (Pezzulo and Rigoli 2011, Solway and Botvinick 2012) or optimal future trajectories (Levine 2018). For example, it is possible to clamp (i.e., fix the value of) the future desired state in the model and then infer the sequence of actions that is more likely to fill the gap from the current state to the future desired state.

Active Inference, planning-as-inference, and other related schemes use a prospective form of control, which starts from an explicit representation of future, to-be-observed states rather than from a set of stimulus-response rules or policies, as is more typically done in optimal control theory and RL. However, the specific implementations of control- and planning-as-inference vary along at least three dimensions—namely, what form of inference they use (e.g., sampling or variational inference), what they infer (e.g., a posterior distribution over actions or action sequences), and the goal of inference (e.g., maximizing the marginal likelihood of an optimality condition or the probability of getting reward).

Active Inference takes a unique perspective on each of these dimensions. First, it uses a scalable approximate scheme—variational inference—to solve the challenging computational problems that arise during planning-as-inference. Second, it affords model-based planning, or the inference of a posterior over control states—which correspond to action sequences or policies, not single actions.[2] Third, to infer action sequences, Active Inference considers the *expected free energy* functional, which mathematically subsumes other widely used planning-as-inference schemes (e.g., KL control) and can handle ambiguous situations (Friston, Rigoli et al. 2015).

10.8   Behavior and Bounded Rationality

> The wise are instructed by reason, average minds by experience, the stupid by necessity and the brute by instinct.
> —Marcus Tullius Cicero

Behavior in Active Inference automatically combines multiple components: deliberative, perseverative, and habitual (Parr 2020). Imagine a person who is walking to a shop close to her house. If she predicts the consequences of her actions (e.g., turning left or right), she can elaborate a good plan to reach the shop. This deliberative aspect of behavior is provided by *expected free energy*, which is minimized when one acts in a way to achieve preferred observations (e.g., being in the shop). Note that *expected free energy* also includes a drive to reduce uncertainty, which can manifest in deliberation. For example, if the person is unsure about the best direction, she can move to an appropriate vantage point, from which she can find the way to the shop easily, even if this implies a longer route. In short, her plans acquire epistemic affordance.

If the person is less able to engage in deliberation (e.g., because she is distracted), she may continue walking after reaching the shop. This perseverative aspect of behavior is provided by *variational free energy*, which is minimized when one gathers observations that are compatible with current beliefs, including beliefs about the current course of actions. The sensory and proprioceptive observations that the person gathers provide evidence for "walking" and hence may determine perseveration in the absence of deliberation.

Finally, another thing the person could do—when she is less able to deliberate—is select the usual plan to go home, without thinking about it. This habitual component is provided by the prior value of policies. This could allocate high probability to a plan to go home—a plan she has observed herself enacting multiple times in the past—and can become dominant if not superseded by deliberation.

Note that deliberative, perseverative, and habitual aspects of behavior coexist and can be combined in Active Inference. In other words, one can infer that, in this situation, a habit is the most likely course of action. This is different from "dual theories," which assume that we are driven by two separate systems, one rational and one intuitive (Kahneman 2017). The mixture of deliberative, perseverative, and habitual aspects of behavior plausibly depends on contextual conditions, such as the amount of experience and the amount of cognitive resources one can invest in deliberative processes that may have a high complexity cost.[3]

The impact of cognitive resources on decision-making has been widely studied under the rubric of bounded rationality (Simon 1990). The core idea is that while an ideal rational agent should always fully consider the outcomes of its actions, a bounded rational agent has to balance the costs, effort, and timeliness of computation—for example, the information-processing costs of deliberating the best plan (Todorov 2009, Gershman et al. 2015).

10.8.1   Free Energy Theory of Bounded Rationality

Bounded rationality has been cast in terms of *Helmholtz free energy minimization*: a thermodynamic construct that is strictly related to the notion of variational free energy as used in Active Inference; see Gottwald and Braun (2020) for details. The "free energy theory of bounded rationality" formulates the trade-offs of action selection with limited information-processing capabilities in terms of two components of free energy: energy and entropy (see chapter 2). The former represents the expected value of a choice (an *accuracy* term), and the latter represents the costs of deliberation (a *complexity* term). What is costly during deliberation is decreasing the entropy (or complexity) of one's beliefs before a choice to render them more precise (Ortega and Braun 2013, Zénon et al. 2019). Intuitively, the choice would be more accurate (and potentially entail higher utility) with a more precise posterior belief, but because increasing the precision of beliefs has a cost, a bounded decision-maker has to find a compromise—by minimizing free energy. The same trade-offs emerge in Active Inference, thus producing forms of bounded rationality. The notion of bounded rationality also resonates with the use of a variational bound on evidence (or marginal likelihood) that is a definitive aspect of Active Inference. In sum, Active Inference provides a model of (bounded) rationality and optimality, where the best solution to a given problem results from the compromise between complementary objectives: accuracy and complexity. These objectives stem from a normative (free energy minimization) imperative that is richer than classical objectives (e.g., utility maximization) usually considered in economic theory.

10.9   Valence, Emotion, and Motivation

> Consider your origins: you were not made to live as brutes, but to follow virtue and knowledge.
> —Dante Alighieri

Active Inference focuses on (negative) free energy as a measure of fitness and the capacity of an organism to realize its goals. While Active Inference proposes that creatures act to minimize their free energy, this does not mean that they ever have to compute it. Generally, it is sufficient to deal with the gradients of the free energy. By analogy, we do not need to know our altitude to find the top of a hill but can simply follow the slope upward. However, some have suggested creatures may model how their free energy changes over time. Proponents of this hypothesis suggest that it might permit characterizations of phenomena like valence, emotion, and motivation.

On this view, it has been proposed that emotional valence, or the positive or negative character of emotions, can be conceived as the rate of change (first time-derivative) of free energy over time (Joffily and Coricelli 2013). Specifically, when a creature experiences an increase in its free energy over time, it may assign a negative valence to the situation; whereas when it experiences a decrease of its free energy over time, it may assign it a positive valence. Extending this line of thought to long-term dynamics of free energy (and second time-derivatives), it may be possible to characterize sophisticated emotional states; for example, the *relief* of passing from a phase of low valence to a phase of high valence, or the *disappointment* of passing from a phase of high valence to a phase of low valence. Monitoring free energy dynamics (and the emotional states they elicit) may permit adapting the behavioral strategies or learning rates to long-term environmental statistics.

It may seem a bit of a leap to assume a second generative model whose role is to monitor the free energy of the first. However, there is another way in which these ideas can be interpreted. An interesting formalization of these perspectives rests on thinking about what causes rapid changes in free energy. As it is a functional of beliefs, a rapid change in free energy must be due to fast belief updating. The key determinant of this speed is precision, which acts as a time-constant in the dynamics of predictive coding. Interestingly, this ties in with the notion of higher derivatives of the free energy, as precision is the negative of the second derivative (i.e., the curvature of a free energy landscape). However, this begs the question as to why we should associate precision with valence. The answer comes from noticing that precision is inversely related to ambiguity. The more precise something is, the less ambiguous its interpretation. Choosing a course of action that minimizes expected free energy also means minimizing ambiguity and therefore maximizing precision. Here we see a direct association between high order derivatives of the free energy, its rate of change, and motivated behavior.

Expectations about (increases or decreases of) free energy may play motivational roles and incentivize behavior, too. In Active Inference, a surrogate expectation about changes (increases or decreases) of free energy is the precision of beliefs about policies. This again highlights the importance of this second order statistic. For example, a highly precise belief signals that one has found a good policy—that is, a policy that can be confidently expected to minimize free energy. Interestingly, the precision of (beliefs about) policies has been linked to dopamine signaling (FitzGerald, Dolan, and Friston 2015). From this perspective, stimuli that increase the precision of beliefs about policies trigger dopamine bursts—which may indicate their incentive salience (Berridge 2007). This perspective may help shed light on the neurophysiological mechanisms linking expectations of goal or reward achievement to increases in attention (Anderson et al. 2011) and motivation (Berridge and Kringelbach 2011).

10.10   Homeostasis, Allostasis, and Interoceptive Processing

There is more wisdom in your body than in your deepest philosophy.
—Friedrich Nietzsche

A creature's generative model is not just about the external world but also—and perhaps even more importantly—about the internal milieu. A generative model of a body's inside (or interoceptive schema) has a dual role: to explain how interoceptive (bodily) sensations are generated and to ensure the correct regulation of physiological parameters (Iodice et al. 2019), like body temperature or sugar levels in the blood. Cybernetic theories (touched on in section 10.6.2) assume that a central objective of living organisms is maintaining homeostasis (Cannon 1929)—ensuring that physiological parameters remain within viable ranges (e.g., body temperature never becomes too high)—and that homeostasis can only be achieved by exerting a successful control over the environment (Ashby 1952).

This form of homeostatic regulation can be achieved in Active Inference by specifying the viable ranges of physiological parameters as priors over interoceptive observations. Interestingly, homeostatic regulation can be achieved in multiple, nested ways. The simplest regulatory loop is the engagement of autonomic reflexes (e.g., vasodilation), when certain parameters are (expected to be) out of range—for example, when body temperature is too high. This autonomic control can be constructed as *interoceptive inference*: an Active Inference process that operates on interoceptive streams rather than proprioceptive streams, as in the case of externally directed actions (Seth et al. 2012, Seth and Friston 2016, Allen et al. 2019). For this, the brain may use a generative model that predicts interoceptive and physiological streams and triggers autonomic reflexes to correct interoceptive prediction errors (e.g., a surprisingly high body temperature). This is analogous to the way motor reflexes are activated to correct proprioceptive prediction errors and steer externally directed actions.

Active Inference extends beyond simple autonomic loops: it can correct the same interoceptive prediction error (high body temperature) in increasingly sophisticated ways (Pezzulo, Rigoli, and Friston 2015). It can use predictive, *allostatic* strategies (Sterling 2012, Barrett and Simmons 2015, Corcoran et al. 2020) that go beyond homeostasis and preemptively control physiology in an allostatic fashion before interoceptive prediction errors are triggered—for example, finding shade before overheating. Another predictive strategy entails mobilizing resources before expected excursions from physiological setpoints—for example, increasing cardiac output before a long run in anticipation of increased oxygen demands. That requires modifying the priors over interoceptive observations dynamically, going beyond homeostasis (Tschantz et al. 2021). Eventually, predictive brains can develop sophisticated goal-directed strategies, such as ensuring that one brings cold water to the beach, meeting the same imperative (controlling body temperature) in richer and more effective ways.

Biological and interoceptive regulation may be crucial for affect and emotional processing (Barrett 2017). During situated interactions, the brain's generative model constantly predicts not just what will happen next but also what the consequences for interoception and allostasis are. Interoceptive streams—elicited during the perception of external objects and events—imbue them with an affective dimension, which signals how good or bad they are for the creature's allostasis and survival, hence making them "meaningful." If this view is correct, then disorders of this interoceptive and allostatic processing may engender emotional dysregulation and various psychopathological conditions (Pezzulo 2013; Barrett et al. 2016; Maisto, Barca et al. 2019; Pezzulo, Maisto et al. 2019).

There is an emerging bedfellow for *interoceptive inference*—namely, *emotional inference*. In this application of Active Inference, emotions are considered part of the generative model: they are just another construct or hypothesis that the brain employs to deploy precision in deep generative models. From the perspective of belief updating, this means anxiety is just a commitment to the Bayesian belief "I am anxious" that best explains the prevailing sensory and interoceptive queues. From the perspective of acting, the ensuing (interoceptive) predictions augment or attenuate various precisions (i.e., covert action) or enslave autonomic responses (i.e., overt action). This may look much like arousal, which confirm the hypothesis that "I am anxious." Usually, emotional inference entails belief updating that is domain general, assimilating information from both interoceptive and exteroceptive sensory streams—hence the intimate relationship between emotion, interoception, and attention in health (Seth and Friston 2016; Smith, Lane et al. 2019; Smith, Parr, and Friston 2019) and disease (Peters et al. 2017, J. E. Clark et al. 2018).

10.11 Attention, Salience, and Epistemic Dynamics

> True ignorance is not the absence of knowledge, but the refusal to acquire it.
> —Karl Popper

Given the number of times we have referred to *precision* and *expected free energy* in this chapter alone, it would be negligent not to devote a little space to *attention* and *salience*. These concepts recur throughout psychology, having been subject to numerous redefinitions and classifications. Sometimes these terms are used to refer to synaptic gain control mechanisms (Hillyard et al. 1998), which preferentially select some sensory modality or subset of channels within a modality. Sometimes they refer to how we orient ourselves, through overt or covert action, to gain more information about the world (Rizzolatti et al. 1987; Sheliga et al. 1994, 1995).

Although the uncertainty afforded by the many meanings of *attention* underwrites some of the epistemic attractiveness of this field of study, there is also value in resolving the attendant ambiguity. One of the things offered by a formal perspective on psychology is that we do not need to worry about this ambiguity. We can operationally define *attention* as the precision associated with some sensory input. This neatly maps to the concept of gain control, as sensations we infer to be more precise will have greater influence over belief updating than those inferred to be imprecise. The construct validity of this association has been demonstrated in relation to psychological paradigms, including the famous Posner paradigm (Feldman and Friston

2010). Specifically, responding to a stimulus at a location in visual space that is afforded a higher precision is faster than responding to stimuli in other locations.

This leaves the term *salience* in want of a similar formal definition. Typically, in Active Inference, we associate salience with expected information gain (or epistemic value): a component of the expected free energy. Intuitively, something is more salient when we expect it to yield more information. However, this defines *salience* in terms of an action or policy, while *attention* is an attribute of beliefs about sensory input. This fits with the notion of salience as overt or covert orienting. We saw in chapter 7 that we could further subdivide expected information gain into salience and novelty. The former is the potential to infer, while the latter is the potential to learn. An analogy that expresses the difference between attention and salience (or novelty) is the design and analysis of a scientific experiment. Attention is the process of selecting the highest quality data from what we have already measured and using these to inform our hypothesis testing. Salience is the design of the next experiment to ensure the highest quality data.

We do not dwell on this issue to simply add another reclassification of attentional phenomena to the literature but to highlight an important advantage in committing to a formal psychology. Under Active Inference, it does not matter if others define *attention* (or any other construct) differently—as we can simply refer to the mathematical constructs in question and preclude any confusion. A final point of consideration is that these definitions offer a simple explanation for why attention and salience are so often conflated. Highly precise data are minimally ambiguous. This means that they should be afforded attention and that actions to acquire these data are highly salient (Parr and Friston 2019a).

10.12 Rule Learning, Causal Inference, and Fast Generalization

> Yesterday I was clever, so I wanted to change the world. Today I am wise, so I am changing myself.
> —Rumi

Humans and other animals excel at making sophisticated causal inferences, learning abstract concepts and the causal relationships between objects, and generalizing from limited experience—in contrast to current machine learning paradigms, which require a large number of examples to attain similar performance. This difference suggests that current machine learning approaches, which are largely based on sophisticated pattern recognition, may not fully capture the ways humans learn and think (Lake et al. 2017).

The learning paradigm of Active Inference is based on the development of generative models that capture the causal relations between actions, events, and observations. In this book, we have considered relatively simple tasks (e.g., the T-maze example of chapter 7) that require unsophisticated generative models. In contrast, understanding and reasoning about complex situations require deep generative models that capture the *latent structure* of the environment—such as hidden regularities that permit generalizing across a number of apparently dissimilar situations (Tervo et al. 2016; Friston, Lin et al. 2017).

One simple example of a hidden rule that governs sophisticated social interactions is a traffic intersection. Imagine a naïve person who observes a busy crossroad and has to predict (or explain) on which occasions pedestrians or cars cross the road. The person can accumulate statistics about the co-occurrence of events (e.g., a red car stopping and a tall man crossing; an old woman stopping and a big car passing), but most are ultimately useless. The person can eventually discover some recurrent statistical patterns, such as that pedestrians cross the road soon after all cars stop at a certain point on the road. This determination would be deemed sufficient in a machine learning setting if the task were just to predict when pedestrians are about to walk, but it would not entail any understanding of the situation. In fact, it may even lead to the erroneous conclusion that the stopping of cars explains the movement of pedestrians. This sort of error is typical in machine learning applications that do not appeal to (causal) models—and cannot distinguish whether the rain explains the wet grass or the wet grass explains the rain (Pearl and Mackenzie 2018).

On the other hand, inferring the correct hidden (e.g., traffic light) rule provides a deeper understanding of the causal structure of the situation (e.g., it is the traffic light that causes the cars to stop and the pedestrians to walk). The hidden rule not only affords better predictive power but also renders inference more parsimonious, as it can abstract away from most sensory details (e.g., the color of cars). In turn, this permits generalizing to other situations, such as different crossroads or cities, where most sensory details differ significantly—with the caveat that facing crossroads in some cities, like Rome, may require more than looking at traffic lights. Finally, learning about traffic light rules may also enable more efficient learning in novel situations—or to develop what is called a "learning set" in psychology or a *learning-to-learn* ability in machine learning (Harlow 1949). When facing a crossroad where the traffic light is off, one cannot use the learned rule but may nevertheless have the expectation that there is another, similar hidden rule in play—and this could help understanding what the traffic police officer is doing.

As this simple example illustrates, learning rich generative models—of the latent structure of the environment (aka structure learning)—affords sophisticated forms of causal reasoning and generalization. Scaling up generative models to address these sophisticated situations is an ongoing objective in computational modeling and cognitive science (Tenenbaum et al. 2006, Kemp and Tenenbaum 2008). Interestingly, there is a tension between current machine learning trends—wherein the general idea is "the bigger, the better"—and the statistical approach of Active Inference—which suggests the importance of balancing the *accuracy* of a model with its *complexity* and to favor simpler models. Model reduction (and the pruning of unnecessary parameters) is not simply a way to avoid wasting resources—it is also an effective way to learn hidden rules, including during offline periods like sleep (Friston, Lin et al. 2017), perhaps manifesting in resting state activity (Pezzulo, Zorzi, and Corbetta 2020).

10.13   Active Inference and Other Fields: Open Directions

> It has to start somewhere, it has to start sometime,
> what better place than here? What better time than now?
> —Rage Against the Machine, "Guerrilla Radio"

In this book, we mainly focus on Active Inference models that address biological problems of survival and adaptation. Yet Active Inference can be applied in many other domains. In this last section, we briefly discuss two such domains: social and cultural dynamics and machine learning and robotics. Addressing the former requires thinking about the ways in which multiple Active Inference agents interact and the emergent effects of such interaction. Addressing the latter requires understanding how Active Inference can be endowed with more effective learning (and inference) mechanisms to scale up to more complex problems—but in a way that is compatible with the basic assumptions of the theory. Both are interesting open directions for research.

10.13.1 Social and Cultural Dynamics

Many interesting aspects of our (human) cognition relate to social and cultural dynamics rather than individualistic perceptions, decisions, and actions (Veissière et al. 2020). By definition, social dynamics require multiple Active Inference creatures that engage in physical interactions (e.g., joint actions, such as playing team sports) or more abstract interactions (e.g., elections or social networking). Simple demonstrations of inter-Active Inference between identical organisms already produced interesting emergent phenomena, such as the self-organization of simple life forms that resist dispersion, the possibility to engage in morphogenetic processes to acquire and restore a body form, and mutual coordinated prediction and turn taking (Friston 2013; Friston and Frith 2015a; Friston, Levin et al. 2015). Other simulations have addressed the ways in which creatures can extend their cognition to material artifacts and shape their cognitive niches (Bruineberg et al. 2018).

These simulations capture only a fraction of the complexity of our social and cultural dynamics, but they illustrate the potential of Active Inference to expand from a science of individuals to a science of societies—and how cognition extends beyond our skulls (Nave et al. 2020).

10.13.2 Machine Learning and Robotics

The generative modeling and variational inference methods discussed in this book are widely used in machine learning and robotics. In these fields, the emphasis is often on how to learn (connectionist) generative models— as opposed to how to use them for Active Inference, the focus of this book. This is interesting as machine learning approaches are potentially useful to scale up the complexity of the generative models and of the problems considered in this book—with the caveat that they may call on very different process theories of Active Inference.

While it is impossible to review here the vast literature on generative modeling in machine learning, we briefly mention some of the most popular models, from which many variants have been developed. Two early connectionist generative models, the *Helmholtz machine* and the *Boltzmann machine* (Ackley et al. 1985, Dayan et al. 1995), provided paradigmatic examples of how to learn the internal representations of a neural network in an *unsupervised* way. The Helmholtz machine is especially related to the variational approach of Active Inference, as it uses separate *recognition* and *generative* networks to infer a distribution over hidden variables and sample from them to obtain fictive data. The early practical success of these methods was limited. But afterward, the possibility to stack multiple (restricted) Boltzmann machines enabled learning of multiple layers of internal representations and was one of the early successes of unsupervised deep neural networks (Hinton 2007).

Two recent examples of connectionist generative models, *variational autoencoders* or VAEs (Kingma and Welling 2014) and *generative adversarial networks* or GANs (Goodfellow et al. 2014), are widely used in machine learning applications, such as recognizing or generating pictures and videos. VAEs exemplify an elegant application of variational methods to learning in generative networks. Their learning objective, the evidence lower bound (ELBO), is mathematically equivalent to variational free energy. This objective enables learning of an accurate description of the data (i.e., maximizes accuracy) but also favors internal representations that do not differ too much from their priors (i.e., minimizes complexity). The latter objective acts as a so-called regularizer, which helps to generalize and avoid overfitting.

GANs follow a different approach: they combine two networks, a *generative network* and a *discriminative network*, which continuously compete during learning. The discriminative network learns to distinguish which example data produced by the generative network are *real* or *fictive*. The generative network tries to generate fictive data that fool (i.e., are misclassified by) the discriminative network. The race between these two networks forces the generative network to improve its generative capabilities and produce high fidelity fictive data—an ability that has been widely exploited to generate, for example, realistic images.

The above generative models (and others) can be used for control tasks. For example, Ha and Eck (2017) have used a (sequence-to-sequence) VAE to learn to predict pencil strokes. By sampling from the internal representation of the VAE, the model can construct novel stroke-based drawings. Generative modeling approaches have been used to control robot movements, too. Some of these approaches use Active Inference (Pio-Lopez et al. 2016, Sancaktar et al. 2020, Ciria et al. 2021) or closely related ideas, but in a connectionist setting (Ahmadi and Tani 2019, Tani and White 2020).

One of the main challenges in this domain is that robot movements are high dimensional and require (learning) sophisticated generative models. One interesting aspect of Active Inference and related approaches is that the most important thing to be learned is a forward mapping between actions and sensory (e.g., visual and proprioceptive) feedback at the next time step. This forward mapping can be learned in various ways: by autonomous exploration, by demonstration, or even by direct interaction with a human—for example, a teacher (the experimenter) who guides the hands of the robot along a trajectory to the goal, hence scaffolding the acquisition of effective goal-directed actions (Yamashita and Tani 2008). The possibility to learn generative models in various ways greatly expands the scope of robot skills that can be eventually achieved. In turn, the possibility to develop more advanced (neuro-) robots using Active Inference could be important not just for technological but also for theoretical reasons. Indeed, some key aspects of Active Inference, such as the adaptive agent-environment interactions, the integration of cognitive functions, and the importance of embodiment, are naturally addressed in robotic settings.

10.14 Summary

Home is behind, the world ahead,
and there are many paths to tread
through shadows to the edge of night,
until the stars are all alight.
—J. R. R. Tolkien, *The Lord of the Rings*

We started this book by asking whether it is possible to understand brain and behavior from first principles. We then introduced Active Inference as a candidate theory to meet this challenge. We hope that the reader has been convinced that the answer to our original question is yes. In this chapter, we considered the unified perspective that Active Inference offers on sentient behavior and what implications this theory has for familiar psychological constructs, such as perception, action selection, and emotion. This gave us the opportunity to revisit the concepts introduced throughout the book and to remind ourselves of the fascinating questions still open for future research. We hope this book provides a useful complement to related works on Active Inference, including on the one hand the philosophy (Hohwy 2013, Clark 2015) and on the other hand the physics (Friston 2019a).

We are now at the end of our journey. Our aim has been to offer an introduction to those interested in using these methods—both at conceptual and formal levels. However, it is important to emphasize that Active Inference is not something that can be learned purely in theory. We encourage anyone who has enjoyed this book to think about pursuing it in practice. Important rites of passage in theoretical neurobiology are trying to write down a generative model, experiencing the frustration when simulations misbehave, and learning from violations of your prior beliefs when something unexpected happens. Whether or not you choose to pursue this practice at a computational level, we hope that you will reflect on it as you engage in Active Inference in day-to-day life. This may manifest in the compulsion to direct your eyes to resolve uncertainty about something in your peripheral vision. It may be in choosing to eat at a favorite restaurant to fulfill prior (gustatory) preferences. It may be in reducing the heat when the shower is too hot to ensure the temperature conforms to your model of how the world should be. Ultimately, we are confident that you will continue to pursue Active Inference in some form.

Appendix A: Mathematical Background

A.1 Introduction

This appendix offers an introduction (or a refresher) to the basic mathematical techniques employed throughout this book. We provide an introductory (but nonexhaustive) overview of four topics: linear algebra, Taylor series approximation, variational calculus, and stochastic dynamics. For each of these techniques, we refer to where it comes into play in the book. Our aim here is to provide a focused introduction—with emphasis on building intuition as opposed to formal and rigorous proofs. The maths required to understand and use Active Inference is not complicated, but its multidisciplinary basis means it is often difficult to find resources that bring together the necessary prerequisites. We hope this appendix goes some way toward remedying this.

A.2 Linear Algebra

A.2.1 The Basics

Linear algebra refers to a notation used to simply and concisely express combinations of multiplications and summations. It relies on matrices and vectors comprising arrays of numbers in structures with multiple rows and columns (or multiple rows and a single column, for a vector). The element of a matrix A in the ith row and jth column is referred to as A_{ij}. The product A of two matrices B and C (or a matrix and vector) is defined as follows:

$$A = BC$$
$$\Rightarrow \tag{A.1}$$
$$A_{ij} = \sum_k B_{ik} C_{kj}$$

For this definition to hold, we need the number of columns of B to match the number of rows of C. However, let us instead say that the number of rows of B match the rows in C and we want to express the following sum:

$$A_{ij} = \sum_k B_{ki} C_{kj} \tag{A.2}$$

How would we do this using linear algebraic notation? We need to appeal to another operation that swaps the subscripted indices of B (i.e., reflects the array such that the columns become rows and vice versa). This is the transpose operation, normally expressed using a superscript T:

$$B_{ik}^T \triangleq B_{ki}$$
$$A = B^T C \triangleq B \cdot C$$
$$\Rightarrow \tag{A.3}$$
$$A_{ij} = \sum_k B_{ki} C_{kj}$$

Equation A.3 shows how we can use the transpose operator to express the summation from equation A.2. The second line highlights an alternative notation using a dot operator. This notation is inspired by the fact that, when B and C have only one column each, equation A.3 reduces to a vector dot product.

Another useful operation is the *trace* operator. This takes the elements along the diagonal of a square matrix and sums them:

$$tr[A] \triangleq \sum_i A_{ii} \tag{A.4}$$

Part of the utility of a trace operator is afforded by the way we can permute elements in the trace of a matrix product:

$$tr[ABC] = \sum_i \sum_j \sum_k A_{ij} B_{jk} C_{ki}$$
$$= \sum_k \sum_i \sum_j C_{ki} A_{ij} B_{jk} = tr[CAB] \tag{A.5}$$
$$= \sum_j \sum_k \sum_i B_{jk} C_{ki} A_{ij} = tr[BCA]$$

The main use we will find for this identity in this book is when applied to scalar quantities. A scalar can be viewed as a matrix with only one row and one column. As such, we can apply a trace operator to it, but this will not do anything—we get the same scalar out. This means that, if a matrix product gives rise to a scalar quantity, we can permute the terms as above.

For example, if we have a square matrix B with N columns and rows, and a vector c with N rows, we can use equation A.5 to show the following:

$$
\begin{aligned}
a &= c \cdot Bc \\
&= tr\left[c^T Bc\right] \\
&= tr\left[Bcc^T\right] \\
&= tr\left[BC\right] \\
C &= c \otimes c \triangleq cc^T
\end{aligned}
\tag{A.6}
$$

This reexpresses a quadratic expression (first line) with the trace of the product of two matrices (penultimate line). The final line defines the outer product (in contrast to the inner dot product). Equation A.6 becomes particularly useful in the context of multivariate normal distributions, as we will come to in section A.2.3.

The final concepts of linear algebra to be aware of are the inverse and determinant of a matrix. An inverse is defined as follows:

$$
A^{-1}A = AA^{-1} = I
\tag{A.7}
$$

Equation A.7 says that the product of a matrix and its inverse is the identity matrix—a square matrix with ones along its main and zeros elsewhere. Multiplying any matrix by the identity matrix returns the original matrix, unchanged. It is the linear algebraic equivalent of scalar multiplication by 1 (which could be interpreted as a 1-dimensional identity matrix). This means that if we multiply something by a matrix, and then by the inverse of that matrix, we end up with the original quantity.

The determinant is a useful quantity but one for which it is harder to develop a clear intuition. The only point at which it appears in this book is as part of the normalizing constant of a multivariate normal distribution. As such, it is worth knowing how it is calculated, but we will not dwell on this concept. The determinant is defined recursively as follows:

$$
|A| \triangleq \sum_i (-1)^{i-1} A_{1i} \left| A_{\setminus(1, i)} \right|
\tag{A.8}
$$

Here, the notation $A_{\setminus(1, i)}$ means the matrix A with row 1 and column i omitted. For example:

$$A = \begin{bmatrix} A_{11} & A_{12} \\ A_{21} & A_{22} \end{bmatrix}$$

$$A_{\backslash(1,1)} = A_{22}$$

$$A_{\backslash(1,2)} = A_{21}$$

$$|A| = A_{11} |A_{22}| - A_{12} |A_{21}|$$

$$= A_{11}A_{22} - A_{12}A_{21}$$

(A.9)

This concludes our outline of the basic operations of linear algebra.

A.2.2 Derivatives

Differentiation of matrix and vector quantities follows directly from the application of standard calculus to each element of a matrix. For example, if we have a matrix B whose elements are functions of a scalar x, the derivative of B with respect to x is as follows:

$$A(x) = \partial_x B(x)$$

$$\Rightarrow A(x)_{ij} = \partial_x B(x)_{ij}$$

$$\partial_x \triangleq \frac{\partial}{\partial x}$$

(A.10)

However, a few important definitions and identities will be useful in understanding the technical details in this book. The first is how to take derivatives with respect to nonscalar quantities. If we have a vector quantity b that is a function of another vector c, the derivative of b with respect to c is a matrix:

$$A = \partial_c b(c)$$

$$\Rightarrow A_{ij} = \partial_{c_j} b(c)_i$$

(A.11)

We will also make use of the gradient operator, which deals with derivatives with respect to a vector. This is defined as follows:

$$\nabla_b = \begin{bmatrix} \partial_{b_1} & \partial_{b_2} & \partial_{b_3} & \cdots \end{bmatrix}^T$$

$$a = \nabla_b x(b)$$

$$\Rightarrow$$

$$a_i = \partial_{b_i} x(b)$$

(A.12)

The definition of the gradient operator as a vector of derivative operators also affords a concise definition of a related quantity—the divergence of a vector function:

$$\nabla_a \cdot b(a) = \sum_i \partial_{a_i} b(a)_i \tag{A.13}$$

There are many useful derivative identities for linear algebraic quantities, but we will not attempt to provide a comprehensive overview; for readers who wish to delve further, we recommend *The Matrix Cookbook* (Petersen and Pedersen 2012). Here, we limit ourselves to two identities that will be particularly useful. The first is the gradient of a quadratic quantity:

$$d(a) = \nabla_a \big(b(a) \cdot Cb(a) \big)$$
$$\Rightarrow$$
$$\begin{aligned} d(a)_i &= \partial_{a_i} \sum_j \sum_k b(a)_j C_{jk} b(a)_k \\ &= \sum_j \sum_k \big(\big(\partial_{a_i} b(a)_j \big) C_{jk} b(a)_k + \big(\partial_{a_i} b(a)_k \big) C_{jk} b(a)_j \big) \end{aligned} \tag{A.14}$$
$$\Rightarrow$$
$$d(a) = \nabla_a b(a) \cdot \big(C + C^T \big) b(a)$$

Here (and throughout this book), the transposition implied by the dot notation is applied prior to the gradient operator:

$$\nabla_a b(a) \cdot (\cdots) \triangleq \nabla_a b(a)^T (\cdots) \neq \big(\nabla_a b(a) \big)^T (\cdots) \tag{A.15}$$

The identity in equation A.14 is used in the derivation of the belief-update equations for predictive coding in chapter 4. A second useful identity is the derivative of the same quantity with respect to the matrix, C:

$$D(a) = \nabla_C \big(b(a) \cdot Cb(a) \big)$$
$$\Rightarrow$$
$$D(a)_{ij} = \partial_{C_{ij}} \sum_k \sum_l b(a)_k C_{kl} b(a)_l = b(a)_i b(a)_j \tag{A.16}$$
$$\Rightarrow$$
$$D(a) = b(a) \otimes b(a)$$

Here we have used the gradient operator with a matrix subscript to indicate the following:

$$\nabla_C = \begin{bmatrix} \partial_{C_{11}} & \partial_{C_{12}} & \cdots \\ \partial_{C_{21}} & \partial_{C_{22}} & \\ \vdots & & \ddots \end{bmatrix} \tag{A.17}$$

In appendix B, we will see how equation A.16 aids in the estimation of the covariance matrix for a posterior probability.

A.2.3 Probabilities

In the context of probabilistic reasoning, these linear algebraic identities come into play in two important situations. The first is when the random variable we are reasoning about (i.e., the support of a probability distribution) is a vector quantity. The second is when the probability distribution itself is described by sufficient statistics that are vectors, matrices, or higher order tensor quantities.[1] An example of both is the multivariate normal distribution, defined as follows:

$$p(x) = \left(\frac{1}{(2\pi)^k} |\Pi| \right)^{\frac{1}{2}} e^{-\frac{1}{2}(x-\eta) \cdot \Pi(x-\eta)}$$
$$\dim(x) = k$$

(A.18)

Here, x is a k-dimensional vector. This means the mode, η, is also a k-dimensional vector. The precision, Π, is the inverse of the covariance—a $k \times k$ dimensional symmetric matrix expressing the dispersion of probability mass around the mode. This appears twice in equation A.18: in the normalizing constant (as a determinant) and in the exponent. Note that the quadratic term in the exponent is a scalar quantity and is therefore susceptible to the identity in equation A.6. This will be important in appendix B.

When dealing with categorical probability distributions, the sufficient statistics of a distribution are simply vectors, matrices, or tensors of probabilities. For example, the probability distribution over the numbers a person could roll on a six-sided die is given by a 6-dimensional vector, with each element of the vector expressing the probability of that number. Things get more interesting in the context of conditional probabilities. For variables o and s, which each take one of several possible values, we can write the conditional probability of o given s as a matrix, A, whose elements are as follows:

$$P(o = i \mid s = j) = A_{ij}$$

(A.19)

This says that the probability that o takes its ith possible value if s takes its jth possible value is given by the element of A in the ith row and jth column. Taking this further, we can define conditional probabilities in which there are multiple items in the conditioning set, leading to a tensor structure:

$$P(o = i \mid s_1 = j, s_2 = k, s_3 = l, \ldots) = A_{ijkl\ldots}$$

(A.20)

Here we could specify an arbitrary number of variables in the conditioning set, leading to an arbitrary number of indices, and a tensor of arbitrary order. We set out an example of a (T-maze) model in chapter 7 that makes use of a probability tensor of order 3. The principles of this model generalize to any higher order. For a tensor A, we will consistently use the dot notation of equation A.3 to mean summation with respect to the first index:

$$A = B \cdot x$$
$$\Rightarrow$$
$$A_{jklm\ldots} = \sum_i B_{ijklm\ldots} x_i$$

(A.21)

An advantage of this expression of distributions as arrays of numbers is that we can use the definitions in sections A.2.1–A.2.2 to find concise expressions for related quantities. For example, we will often need to compute information-theoretic quantities like entropies for probability distributions. An entropy is a negative expected (average) log probability. If we take the expression in equation A.19, we can find a simple form for its entropy as follows:

$$H[P(o|s)] \triangleq -\mathbb{E}_{P(o|s)}[\ln P(o|s)]$$
$$\mathbf{H}_j \triangleq H[P(o|s=j)]$$
$$= -\sum_i P(o=i|s=j)\ln P(o=i|s=j)$$
$$\Rightarrow$$
$$\mathbf{H} = -diag(A \cdot \ln A)$$

(A.22)

In equation A.22, *diag* is an operation that takes the diagonal elements of a matrix and stacks them into a vector. This illustrates an example in chapter 4 of defining the expected free energy, in which an appeal to linear algebraic notation offers a concise description of how these quantities may be calculated.

A.3   Taylor Series Approximation

A.3.1   Introduction
Often, it is convenient to simplify the form of a function ($f(x)$) through an approximation (indicated by ^) that is valid in a local region (e.g., the region around a point, a). If we were only interested in the function at a, we

could replace the function with a constant equal to the function evaluated at that point:

$$\hat{f}(x) = f(a) \tag{A.23}$$

However, this is only valid when x is exactly equal to a. In order to make the approximation valid in the region immediately surrounding a, we can add a term to ensure that a small change in x is accompanied by a change in the value of the function consistent with the gradient at a:

$$\hat{f}(x) = f(a) + \varepsilon \partial_x f(x)\big|_{x=a}$$
$$\varepsilon \triangleq x - a \tag{A.24}$$

When x is equal to a, the ε term is zero, consistent with equation A.23. In addition, the first derivative of the original function and of the approximation are equal, when evaluated at a.

Pursuing this approach, we can add an additional term that accounts for the rate of change of the gradient (i.e., the curvature) so that the approximation becomes valid for a greater deviation from a. We do not have to stop here; we could add an arbitrary number of terms to match each successive derivative between the original function and the approximation:

$$\hat{f}(x) = f(a) + \varepsilon \partial_x f(x)\big|_{x=a} + \frac{1}{2} \varepsilon^2 \partial_x^2 f(x)\big|_{x=a} + \cdots$$
$$= \sum_{n=0} \frac{1}{n!} \varepsilon^n \partial_x^n f(x)\big|_{x=a} \tag{A.25}$$

Equation A.25 shows the Taylor series expansion in one dimension. However, we can generalize this to the multivariate case (where x is a vector) with the following expression:

$$f(x) \approx f(a) + \varepsilon \cdot \nabla_x f(x)\big|_{x=a} + \frac{1}{2} \varepsilon \cdot \nabla_x \left(\nabla_x f(x)\right)^T \big|_{x=a} \varepsilon + \cdots \tag{A.26}$$

The quantity $\nabla_x(\nabla_x f(x))^T$ is known as a Hessian matrix.

Increasing the number of terms in the series improves the approximation. For our purposes, we need not go beyond the second order (quadratic) expansion. In the following subsections, we highlight the places in this book in which this approximation has been exploited. These include the Laplace approximation, which underwrites the predictive coding schemes described in chapters 4 and 8 and the variational Laplace scheme used for model-based data analysis described in chapter 9. In addition, the generalized coordinates of motion used to model continuous trajectories (box 4.2) can be interpreted as Taylor series coefficients. We will unpack these applications in sections A.3.2 and A.3.3, respectively.

A.3.2 The Laplace Approximation

An important application of a Taylor series approximation in probabilistic inference is its use in the Laplace approximation. This refers to the use of a Gaussian distribution to approximate a probability distribution (p) in the region surrounding its mode (μ). If we expand the log of a probability distribution using equation A.26, we get the following:

$$\ln p(x) \approx \ln p(\mu) + \varepsilon \cdot \nabla_x \ln p(x)\big|_{x=\mu} + \frac{1}{2}\varepsilon \cdot \nabla_x \left(\nabla_x \ln p(x)\right)^T \big|_{x=\mu} \varepsilon \tag{A.27}$$

$$\varepsilon \triangleq x - \mu$$

This is simply equation A.26 but with $f(x) = \ln p(x)$ and $a = \mu$. The first term after the approximate equality is constant with respect to x so may be absorbed into a normalizing constant. The second term disappears, as the gradient of the log probability at its mode is zero. Exponentiating both sides leaves us with this:

$$p(x) \approx \frac{1}{Z} e^{-\frac{1}{2}\varepsilon \cdot C^{-1}\varepsilon}$$
$$= \mathcal{N}(\mu, C^{-1}) \tag{A.28}$$
$$C^{-1} \triangleq -\nabla_x \left(\nabla_x \ln p(x)\right)^T \big|_{x=\mu}$$

Equation A.28 says that when we approximate a log probability using a quadratic function, near its mode, the associated probability density is Gaussian. This is the Laplace approximation applied to a probability distribution. However, we can also apply the Laplace approximation to a free energy functional. To provide some intuition for this, we start with a free energy functional (see chapter 4):

$$F[q,y] = \mathbb{E}_{q(x)}[\ln q(x) - \ln p(y,x)] \tag{A.29}$$

Equation A.29 expresses free energy in terms of the expected difference between two log probabilities. The q density is an approximate posterior probability. The p density is a generative model, describing how hidden states

(x) give rise to data (y). As in equation A.27, we can apply a Taylor series expansion to the two log probabilities. Starting with the variational density, we have this:

$$\ln q(x) \approx \ln q(\mu) + (x - \mu) \cdot \underbrace{\nabla_x \ln q(x)\big|_{x=\mu}}_{0}$$

$$+ \tfrac{1}{2}(x - \mu) \cdot \nabla_x \left(\nabla_x \ln q(x)\right)^T\big|_{x=\mu} (x - \mu)$$

$$\Rightarrow q(x) \approx \mathcal{N}\left(\mu, \Sigma^{-1}\right) \tag{A.30}$$

$$\Sigma^{-1} = -\nabla_x \left(\nabla_x \ln q(x)\right)^T\big|_{x=\mu}$$

$$\mu = \arg\max_x q(x)$$

Applying the expectation from equation A.29 to equation A.30, we get this:

$$\mathbb{E}_{q(x)}\left[\ln q(x)\right] \approx \ln q(\mu) - \tfrac{1}{2}\mathbb{E}_{q(x)}\left[(x - \mu) \cdot \Sigma^{-1}(x - \mu)\right]$$

$$= \ln q(\mu) - \tfrac{1}{2}tr\left[\Sigma^{-1}\underbrace{\mathbb{E}_{q(x)}\left[(x - \mu)(x - \mu)^T\right]}_{\Sigma}\right] \tag{A.31}$$

$$= -\tfrac{k}{2}\ln 2\pi - \tfrac{1}{2}\ln|\Sigma| - \tfrac{k}{2}$$

$$= -\tfrac{1}{2}\ln(2\pi e)^k|\Sigma|$$

Here, k is the dimensionality of x. The move from the first to the second line depends on the trace identity in equation A.6. The first two terms in the third line come from the definition of a multivariate normal distribution (equation A.18). Equation A.31 expresses the first term of equation A.27 under the Laplace assumption. The second term of equation A.27 can similarly be expanded around μ:

$$\ln p(y, x) \approx \ln p(y, \mu) + (x - \mu) \cdot \nabla_x \ln p(y, x)\big|_{x=\mu}$$

$$+ \frac{1}{2}(x - \mu) \cdot \nabla_x \left(\nabla_x \ln p(y, x)\right)^T\big|_{x=\mu} (x - \mu)$$

$$\mathbb{E}_{q(x)}\left[\ln p(y, x)\right] \approx \ln p(y, \mu) + \underbrace{(\mu - \mathbb{E}_{q(x)}[x])}_{0}\nabla_x \ln p(y, x)\big|_{x=\mu} \tag{A.32}$$

$$+ \frac{1}{2}tr\left[\mathbb{E}_{q(x)}[(x - \mu)(x - \mu)^T]\nabla_x \left(\nabla_x \ln p(y, x)\right)^T\big|_{x=\mu}\right]$$

$$= \ln p(y, \mu) + \frac{1}{2}tr\left[\Sigma\nabla_x \left(\nabla_x \ln p(y, x)\right)^T\big|_{x=\mu}\right]$$

The final equality uses the fact that, if q is a normal distribution, its mean is also its mode. Substituting equations A.31 and A.32 back into equation A.27, we get the Laplace free energy:

$$F[q,y] \approx -\frac{1}{2}\ln(2\pi e)^k|\Sigma| - \ln p(y,\mu) - \frac{1}{2}tr\left[\Sigma \nabla_x \left(\nabla_x \ln p(y,x)\right)^T \Big|_{x=\mu}\right] \quad \text{(A.33)}$$

The trace operator in the last term can be ignored when x is 1-dimensional. The useful thing about this formulation is that if we set the derivative of the free energy with respect to the posterior covariance to zero, we find the following:[2]

$$\partial_\Sigma F[q,y] = 0 \Leftrightarrow \Sigma^{-1} = -\nabla_x \left(\nabla_x \ln p(y,x)\right)^T \Big|_{x=\mu} \quad \text{(A.34)}$$

This means that the precision of the posterior is the negative curvature of the log probability of states and data evaluated at the posterior mode. As such, minimizing free energy does not require explicit optimization of the precision—this may be computed analytically from the posterior mean. Furthermore, substitution of A.34 into A.33 reveals that the only term in the free energy that depends on the posterior mean is the log probability over data and states. For details of how this is done to perform inference in continuous state-space models, see chapter 4.

A.3.3   Generalized Coordinates of Motion

In addition to being central to the Laplace approximation, the Taylor series approximation plays another important role in Active Inference. This is in the use of generalized coordinates of motion to represent beliefs about a trajectory through time. In brief, this means drawing inferences not only about the position of a variable (x) but also its velocity (x'), acceleration (x''), and subsequent temporal derivatives. These implicitly represent an approximation to the trajectory that can be made explicit through the following Taylor series:

$$x(t) \approx x(\tau) + \varepsilon x'(t)\big|_{t=\tau} + \frac{1}{2}\varepsilon^2 x''(t)\big|_{t=\tau} + \cdots \quad \text{(A.35)}$$

$$\varepsilon = t - \tau$$

This additionally means we can account for structure in the covariance of random fluctuations, as is necessary in dealing with these fluctuations in biological systems (where fluctuations are themselves generated by dynamical processes). We will discuss this further in section A5. For now, we simply note that a probability density over the generalized coordinates of motion is equivalent to a distribution over local trajectories constructed by treating the coordinates as coefficients of a Taylor series expansion.

A.4 Variational Calculus

A.4.1 Functional Derivatives

Because Active Inference deals with optimizing beliefs (probability distributions), it is often necessary to talk about the minimization of functionals (functions of functions) with respect to functions. This calls for the concept of a functional (i.e., variational) derivative. The basic problem is finding the function (f) that minimizes a functional (S), normally expressed as an integral[3] of a function that includes f:

$$\phi(x) = \arg\min_{f} S[f(x)]$$

$$S[f(x)] \triangleq \int_{x_1}^{x_2} \mathcal{L}(f(x), x) \, dx$$

(A.36)

If we parameterize the function in terms of an arbitrary function (g) that is zero at the extremes of the integral and multiply this by a small number (u), we can take the derivative of S with respect to u:

$$f(x,u) \triangleq \phi(x) + u g(x)$$

$$\partial_u S[f(x,u)] = \int_{x_1}^{x_2} \partial_u \mathcal{L}(f(x,u), x) \, dx$$

$$= \int_{x_1}^{x_2} \partial_u f(x,u) \partial_f \mathcal{L}(f(x,u), x) \, dx$$

$$= \int_{x_1}^{x_2} g(x) \partial_f \mathcal{L}(f(x,u), x) \, dx$$

(A.37)

When u is zero, f is the function that minimizes the integral. This means equation A.37 should be zero when evaluated at $u = 0$. The condition that must be satisfied for f to minimize S is then as follows:

$$\int_{x_1}^{x_2} g(x) \partial_f \mathcal{L}(f(x), x) \, dx \Big|_{f=\phi} = 0$$

(A.38)

For equation A.38 to be true for any arbitrary $g(x)$, the following is implied:[4]

$$\delta_f S \triangleq \partial_f \mathcal{L} = 0 \tag{A.39}$$

Note that, in a physics setting, $\mathcal{L}$ may include the gradient of f in addition to the function itself. The same steps outlined above then give rise to the Euler-Lagrange equation:

$$\delta_f S \triangleq \partial_f \mathcal{L} - \frac{d}{dx} \partial_{f'} \mathcal{L} = 0 \tag{A.40}$$

$$f' \triangleq \partial_x f$$

Depending on whether $\mathcal{L}$ includes the gradient, equations A.39 and A.40 express the notion of a variational (aka functional) derivative.

A.4.2 Variational Bayes

Variational Bayes follows in a relatively straightforward way from the above if we set f to be a factor of an approximate posterior distribution and S to be a free energy functional:

$$f(x) = q_i(x_i)$$

$$q(x) = \prod_i q_i(x_i)$$

$$\mathcal{L}(q_i(x_i), x_i) = \int q(x)(\ln q(x) - \ln p(y, x)) dx_{j \neq i} \tag{A.41}$$

$$S[q(x)] = F[q(x), y]$$

The second line here expresses a *mean-field* approximation, in which the approximate posterior is factorized over the variables x. This is often used for reasons of computational tractability. However, this is one of many choices of form for the approximate posterior. Applying equation A.39, we find the form of the approximate posterior that minimizes the free energy (omitting constants):

$$\delta_{q_i} F[q, y] = \ln q_i(x_i) - \int \prod_{j \neq i} q_j(x_j) \ln p(y, x) dx_{j \neq i}$$

$$\delta_{q_i} F[q, y] = 0 \Leftrightarrow \tag{A.42}$$

$$\ln q_i(x_i) = \mathbb{E}_{q \backslash i}[\ln p(y, x)]$$

The notation $\backslash i$ should be read as "all factors except for the ith factor." Equation A.42 is central to an inference scheme known as variational message passing (Winn and Bishop 2005, Dauwels 2007). This works by optimizing each factor of q independently and relies on p being relatively sparse (i.e., not every x_i depends on every other x_j). To gain some intuition for this, consider what happens with an (arbitrary) example:

$$p(y,x) = p(y|x_1)p(x_1|x_2,x_3)p(x_3)p(x_2|x_4)p(x_4)$$
$$\Rightarrow$$
$$\ln q(x_1) =$$
$$= \mathbb{E}_{q(x_2)q(x_3)q(x_4)}\left[\ln p(y|x_1) + \ln p(x_1|x_2,x_3) + \underbrace{\ln p(x_3)p(x_2|x_4)p(x_4)}_{\text{constant w.r.t. } x_1}\right] \quad (A.43)$$
$$\ln q(x_2) =$$
$$= \mathbb{E}_{q(x_1)q(x_3)q(x_4)}\left[\ln p(x_1|x_2,x_3) + \ln p(x_2|x_4) + \underbrace{\ln p(y|x_1)p(x_3)p(x_4)}_{\text{constant w.r.t. } x_2}\right]$$
$$\vdots$$

Equation A43 shows what happens when we substitute the density in the first line into equation A.42 for the first two factors of q. Omitting constant terms, we have this:

$$\ln q(x_1) = \ln p(y|x_1) + \mathbb{E}_{q(x_2)q(x_3)}\left[\ln p(x_1|x_2,x_3)\right]$$
$$\ln q(x_2) = \mathbb{E}_{q(x_1)q(x_3)}\left[\ln p(x_1|x_2,x_3)\right] + \mathbb{E}_{q(x_4)}\left[\ln p(x_2|x_4)\right] \quad (A.44)$$
$$\vdots$$

The terms in the expectation have been simplified by noting the following:

$$\mathbb{E}_{p(b)}\left[f(a)\right] = \int p(b)f(a)\,db = f(a)\underbrace{\int p(b)\,db}_{=1} = f(a) \quad (A.45)$$

This accounts for the simplicity of variational message passing, in which we only need take account of a small subset of beliefs (those about the Markov blanket—see box 4.1) in order to update each belief.

A.5   Stochastic Dynamics

A.5.1   Stochastic Differential Equations

There are a few places in this book where we refer to ideas from the theory of random dynamical systems. In chapter 3, for instance, we highlight the importance of a steady-state distribution to which a random system tends over time and the relationship between these dynamics and the notion of *self-evidencing*. In chapters 4 and 8, we outline how a continuous state-space model may be formulated in terms of stochastic differential equations.

Although this is a fascinating topic (Yuan and Ao 2012), a full dissection of the subtleties of defining stochastic processes is outside the scope of this book. However, it is worth briefly unpacking what we mean by a stochastic differential equation. Put simply, it is a differential equation that is augmented by a random term (ω):

$$\dot{x} = f(x) + \omega$$
$$\omega \sim \mathcal{N}(0, \tfrac{1}{2}\Gamma^{-1}) \tag{A.46}$$

The random term here is chosen to be normally distributed. It has a mean of zero, such that the most likely value for the rate of change of x is simply $f(x)$. The interpretation of equation A.46 is sometimes a little tricky. The best way to dispel any ambiguity is to see it as the limiting case of a discretized scheme:

$$\Delta x = f(x)\Delta\tau + \omega(\Delta\tau)^{\frac{1}{2}}$$
$$\Delta\tau \to 0 \Rightarrow \dot{x} = f(x) + \omega \tag{A.47}$$

Note that if the variance of ω varies with x there are multiple discretizations we could appeal to. The most common choices correspond to Ito and Stratonovich interpretations of a stochastic equation. However, we assume a fixed variance throughout this book—which ensures these interpretations lead to identical results. For the purpose of defining a generative model of the sort found in chapter 8, we just need the probability distribution describing the rate of change of x. From equation A.46, this is simply as follows:

$$p(\dot{x}\,|\,x) = \mathcal{N}(f(x), \tfrac{1}{2}\Gamma^{-1}) \tag{A.48}$$

This is the form that will be found in the generative models used here. This provides a summary of the distinction between a deterministic and a random dynamical system. If we know the value of x in a deterministic system, then we know its velocity. In a stochastic system, knowing x tells us the distribution of possible velocities we might expect.

A.5.2 Nonequilibrium Steady State

In chapter 3, we see that a system defined such that it descends some energy (or surprise) function maintains its form over time and persists at a (possibly nonequilibrium) steady state. We will briefly unpack what this means here, starting from the idea of a steady state and recovering the surprise-minimizing or "self-evidencing" (Hohwy 2016) dynamics. The starting point is an alternative expression of the stochastic dynamics in equation A.46 in terms of a deterministic partial differential equation describing how the probability density changes over time. This is known as a Fokker-Planck equation (Risken 1996):

$$\partial_\tau p(x) = \nabla_x \cdot \left(\Gamma \nabla_x p(x) - f(x) p(x) \right) \tag{A.49}$$

The Fokker-Planck equation lets us define a steady state simply by setting the partial derivative of the density with respect to time to be zero:

$$\partial_\tau p(x) = 0$$
$$\Rightarrow$$
$$\nabla_x \cdot \left(\Gamma \nabla_x p(x) - f(x) p(x) \right) = 0$$
$$\Rightarrow \tag{A.50}$$
$$f(x) = -(\Gamma - Q(x)) \nabla_x \Im(x)$$
$$\nabla_x \cdot \left(Q(x) \nabla_x p(x) \right) = 0$$
$$\Im(x) \triangleq -\ln p(x)$$

The third equality here[5] is key, as it says that those systems that maintain steady state must exhibit dynamics that (on average) minimize their surprise ($\Im$). The Q term allows for dynamics along the contours of the surprise, which neither increase nor decrease surprise. This expression underwrites the self-evidencing perspective of Active Inference and is central to the physics of sentient systems. We will not dwell on this here but refer readers to Friston (2019a) for a more comprehensive overview of the consequences of this treatment.

A.5.3 Generalized Coordinates of Motion

As we saw in section A.3.3, we can represent a short trajectory in terms of the coefficients of a Taylor series expansion in time. This raises an interesting question when we translate this into the context of a stochastic setting. When specifying a continuous-time model in terms of generalized coordinates of motion, how do we account for the covariance between the orders of generalized motion? The answer is given in Cox and Miller (1965), which we summarize here. A random process is expressed in generalized coordinates as a vector of the random fluctuations accompanying the flow, the rate of change of that flow, and subsequent temporal derivatives:

$$\dot{\tilde{x}} = \tilde{f}(\tilde{x}) + \tilde{\omega}$$

$$\tilde{\omega} \triangleq \begin{bmatrix} \omega \\ \omega' \\ \omega'' \\ \omega''' \\ \vdots \end{bmatrix} = \begin{bmatrix} \omega^{[0]} \\ \omega^{[1]} \\ \omega^{[2]} \\ \omega^{[3]} \\ \vdots \end{bmatrix} \tag{A.51}$$

The random fluctuations may be characterized as follows:

$$p(\tilde{\omega}) = \mathcal{N}(0, \tilde{\Pi})$$
$$\mathbb{E}[\omega^{[0]}(\tau)] = 0 \tag{A.52}$$
$$\mathbb{E}[\omega^{[0]}(\tau) \cdot \omega^{[0]}(\tau)] = \Sigma$$

Their autocorrelation function is this:

$$\rho(h) \triangleq \Sigma^{-1} \underbrace{\mathbb{E}[\omega^{[0]}(\tau) \cdot \omega^{[0]}(\tau + h)]}_{\text{Covariance}} \tag{A.53}$$

We can multiply both sides of this equation by the variance to show that the covariance between the noise at two time-points may be factorized into an autocorrelation and a variance. We define the ith derivative of the random fluctuations as this limiting case:

$$\omega^{[i]}(\tau, \Delta\tau) = \frac{\omega^{[i-1]}(\tau + \Delta\tau) - \omega^{[i-1]}(\tau)}{\Delta\tau} \tag{A.54}$$

Using equations A.52 and A.53, we can express the covariance between a variable and its first temporal derivative:

$$\mathbb{E}[\omega^{[1]}(\tau, \Delta\tau) \cdot \omega^{[0]}(\tau + h)] = \frac{1}{\Delta\tau}\mathbb{E}\Big[\big(\omega^{[0]}(\tau + \Delta\tau) - \omega^{[0]}(\tau)\big)\omega^{[0]}(\tau + h)\Big]$$
$$= \frac{1}{\Delta\tau}\Sigma\big(\rho(h - \Delta\tau) - \rho(h)\big) \tag{A.55}$$

Taking the limit as the change in time tends to zero:

$$\mathbb{E}[\omega^{[1]}(\tau) \cdot \omega^{[0]}(\tau + h)] = \Sigma \dot{\rho}(h) \tag{A.56}$$

Evaluating at $h = 0$ gives us a covariance of zero, as the instantaneous velocity and position are orthogonal to one another (and the autocorrelation is at a maximum, so its temporal derivative is zero).

We can take this procedure one step further and evaluate the variance of the first derivative:

$$\mathbb{E}[\omega^{[1]}(\tau,\Delta\tau)\cdot\omega^{[1]}(\tau+h,\Delta\tau)]$$

$$=\tfrac{1}{\Delta\tau^2}\sum\mathbb{E}\Big[\big(\omega^{[0]}(\tau+\Delta\tau)-\omega^{[0]}(\tau)\big)\big(\omega^{[0]}(\tau+h+\Delta\tau)-\omega^{[0]}(\tau+h)\big)\Big] \qquad (A.57)$$

$$=\sum\tfrac{1}{\Delta\tau}\Big(\tfrac{1}{\Delta\tau}\big(\rho(h)-\rho(h-\Delta\tau)\big)-\tfrac{1}{\Delta\tau}\big(\rho(h+\Delta\tau)-\rho(h)\big)\Big)$$

Taking the limit as $\Delta\tau\to 0$, this is as follows:

$$\mathbb{E}[\omega^{[1]}(\tau)\cdot\omega^{[1]}(\tau+h)]=-\sum\ddot{\rho}(h) \qquad (A.58)$$

Pursuing this procedure for subsequent derivatives allows us to compute the elements of the generalized precision matrix:

$$\tilde{\Pi}=\Sigma^{-1}\otimes
\begin{bmatrix}
1 & 0 & \ddot{\rho}(0) & \\
0 & -\ddot{\rho}(0) & 0 & \\
\ddot{\rho}(0) & 0 & \ddddot{\rho}(0) & \\
& & & \ddots
\end{bmatrix}^{-1} \qquad (A.59)$$

Choosing the autocorrelation function to be Gaussian, we have the following:

$$
\begin{aligned}
\rho(h) &= e^{-\frac{1}{2}\lambda h^2} & \rho(0) &= 1 \\
\dot{\rho}(h) &= -\lambda h\rho(h) & \dot{\rho}(0) &= 0 \\
\ddot{\rho}(h) &= \lambda(\lambda h^2-1)\rho(h) & \ddot{\rho}(0) &= -\lambda \\
\dddot{\rho}(h) &= -\lambda^2 h(\lambda h^2-3)\rho(h) & \dddot{\rho}(0) &= 0 \\
\ddddot{\rho}(h) &= \lambda^2(\lambda^2 h^4-6\lambda h^2+3)\rho(h) & \ddddot{\rho}(0) &= 3\lambda^2
\end{aligned}
\qquad (A.60)
$$

The precision term (λ) can then be thought of as parameterizing the smoothness of the random fluctuations. This may itself be optimized in relation to data through minimization of free energy.

Appendix B: The Equations of Active Inference

B.1 Introduction

In this appendix, we provide a mathematical summary of Active Inference. This supplements the equations in the mainchapters with details about where they come from and aims to fill in some of the intermediate steps omitted there. This builds directly on the mathematical background of appendix A and deals with inference in partially observed Markov decision (POMDP) processes and predictive coding architectures, and it touches on questions of structure learning and model reduction alluded to in the main text. Our aim is for this to be relatively self-contained, with particular focus on topics that frequently cause confusion. Readers should be reassured that it is not necessary to understand everything in this appendix to be able to usefully apply Active Inference; this is more for those who want greater technical detail.

B.2 Markov Decision Processes

B.2.1 State Inference

When solving a POMDP problem, our aim is to select the appropriate course of action, or policy. Under Active Inference, this is framed as an inference problem, in which we must find a posterior probability distribution over alternative policies. To calculate a posterior probability, we need two things: the prior probability of policies (addressed in section B.2.2) and the likelihood of observations given a policy. This section focuses on the latter.

The likelihood of observations given a policy is not straightforward to compute. This is because a POMDP problem is structured so that policies (π)

influence trajectories (indicated by ~) of states (s) that influence outcomes (o) without a direct influence of policies on outcomes. The problem then involves a sum over trajectories of states to marginalize these out and find a marginal likelihood of observations given policies:

$$P(\tilde{o}|\pi) = \sum_{\tilde{s}} P(\tilde{o}|\tilde{s})P(\tilde{s}|\pi) \tag{B.1}$$

For any nontrivial state-space, this summation can be very challenging to compute, from a computational perspective. However, as we see in chapter 2, we can approximate marginal likelihoods of this sort using a free energy functional. Chapters 2–4 describe free energy as a functional of two things: approximate posterior beliefs (Q) and a generative model (P). This lets us express the free energy for a given policy as follows:

$$F(\pi) = \mathbb{E}_{Q(\tilde{s}|\pi)}[\ln Q(\tilde{s}|\pi) - \ln P(\tilde{o}, \tilde{s}|\pi)] \geq -\ln P(\tilde{o}|\pi)$$
$$Q(\tilde{s}|\pi) = \arg\min_{Q} F(\pi) \Rightarrow F(\pi) \approx -\ln P(\tilde{o}|\pi) \tag{B.2}$$

Equation B.2 tells us something simple but important. To be able to infer what to do, we need to approximate a marginal likelihood of a policy. To find a good approximation of this marginal likelihood, we need to optimize our beliefs about states under that policy. In short, perceptual inference is mandated for planning to proceed. So how do we solve this problem practically? The answer is to appeal to the methods outlined in section A.4.2. By choosing explicit forms for the probability distributions in equation B.1, we can find a simple expression for the free energy:

$$Q(\tilde{s}|\pi) = \prod_{\tau} Q(s_{\tau}|\pi): \ Q(s_{\tau}|\pi) = Cat(\mathbf{s}_{\pi\tau})$$
$$P(\tilde{o}|\tilde{s}) = \prod_{\tau} P(o_{\tau}|s_{\tau}): \ P(o_{\tau}|s_{\tau}) = Cat(\mathbf{A})$$
$$P(\tilde{s}|\pi) = P(s_1)\prod_{\tau} P(s_{\tau+1}|s_{\tau}, \pi): \ P(s_{\tau+1}|s_{\tau}, \pi) = Cat(\mathbf{B}_{\pi\tau}) \tag{B.3}$$
$$P(s_1) = Cat(\mathbf{D})$$

Briefly, the first line of equation B.3 defines beliefs about states in terms of a mean-field approximation (see equation A.41), factorized over time. Each time-point is associated with a belief about what the state would be on pursuing a policy, given by the vector $\mathbf{s}_{\pi\tau}$, whose elements are the probabilities of each alternative state. The trajectory of observations in the second line depends on a trajectory of hidden states, with the matrix (or tensor, if the states are further factorized) **A** indicating the distribution over observations for each state. Similarly, the prior trajectory of states under a model comprises the transition probabilities under that policy ($\mathbf{B}_{\pi\tau}$) and the initial state probabilities (**D**). Substituting these into the free energy expression in Equation B.2, we arrive at the following expression for the free energy under a policy:

$$\mathbf{F}_{\pi} = \mathbf{s}_{\pi 1} \cdot (\ln \mathbf{s}_{\pi 1} - \ln \mathbf{A} \cdot o_1 - \ln \mathbf{D}) + \sum_{\tau=2} \mathbf{s}_{\pi\tau} \cdot (\ln \mathbf{s}_{\pi\tau} - \ln \mathbf{A} \cdot o_{\tau} - \ln \mathbf{B}_{\pi\tau} \mathbf{s}_{\pi\tau-1}) \text{ (B.4)}$$

Note that the dot product of a probability vector with another quantity is equivalent to the expectation operation. See section A.2.1 if this is not clear. Equation B.4 treats the outcomes as if they were probability vectors, but with a one in the element corresponding to the observed outcome and zeros elsewhere (sometimes called one-hot encoding or 1-in-k vector). The challenge now is to minimize the free energy with respect to our beliefs about states ($\mathbf{s}_{\pi\tau}$) to ensure the free energy becomes a good approximation to a marginal likelihood. We could do this as in section A.4.2 and minimize with respect to each factor of our beliefs one at a time, iterating through until they converge. However, as we are interested in more biologically plausible schemes, we can instead construct a dynamical system that converges on the same solution. This approach is known as a gradient descent, as we follow the free energy gradients downward until we arrive at the minimum.

To update beliefs about states, we take the gradient of this with respect to current beliefs about states. We then define an auxiliary variable (**v**) that plays the role of the log posterior and set this to perform a gradient descent on the free energy. This log posterior is then passed through a softmax function[1] (σ) to convert it to a normalized probability distribution. This process ensures that beliefs about the states change such that they decrease free energy.

$$\mathbf{s}_{\pi\tau} = \sigma(\mathbf{v}_{\pi\tau})$$
$$\dot{\mathbf{v}}_{\pi\tau} = -\nabla_{\mathbf{s}_{\pi\tau}} \mathbf{F}_{\pi} \qquad\qquad\qquad\qquad (B.5)$$
$$\nabla_{\mathbf{s}_{\pi\tau}} \mathbf{F}_{\pi} = \ln \mathbf{s}_{\pi\tau} - \ln \mathbf{A} \cdot o_{\tau} - \ln \mathbf{B}_{\pi\tau} \mathbf{s}_{\pi\tau-1} - \ln \mathbf{B}_{\pi\tau+1} \cdot \mathbf{s}_{\pi\tau+1}$$

Equation B.5 has the same solution to the variational message passing scheme outlined in equation A.42. It allows for efficient computation of posterior beliefs using only locally derived information (in this case, from sensory data, beliefs about the immediate past and beliefs about the immediate future). However, it is worth noting that the mean-field approximation used here (factorization over time) often leads to overconfident posteriors. In practice, this may be countered using a modified scheme called marginal message passing (Friston, FitzGerald et al. 2017; Parr, Markovic et al. 2019):

$$\dot{\mathbf{v}}_{\pi\tau} = \boldsymbol{\varepsilon}_{\pi\tau}$$
$$\boldsymbol{\varepsilon}_{\pi\tau} = \ln \mathbf{A} \cdot o_\tau + \tfrac{1}{2}\left(\ln(\mathbf{B}_{\pi\tau-1}\mathbf{s}_{\pi\tau-1}) + \ln(\mathbf{B}^{\dagger}_{\pi\tau}\mathbf{s}_{\pi\tau+1})\right) - \ln \mathbf{s}_{\pi\tau} \qquad (B.6)$$
$$\mathbf{B}^{\dagger}_{\pi\tau} \propto \mathbf{B}^{T}_{\pi\tau}$$

This leads to more conservative inferences, with greater uncertainty ascribed to posterior beliefs. Other alternatives have been explored, including the Bethe approximation (Schwöbel et al. 2018). However, at the time of writing, the most widely used implementation of Active Inference employs marginal message passing.

B.2.2 Planning as Inference

The above section deals with inference about states conditioned on some policy to minimize a free energy conditioned on the policy. This free energy plays the role of a negative log marginal likelihood (model evidence), wherein each policy is treated as a model. Equipping this with prior and posterior beliefs about the most likely policy, we can express the free energy as a functional of beliefs about policies.

$$\begin{aligned}
F &= \mathbb{E}_{Q(\pi)}[\ln Q(\pi) - \ln P(\pi, \tilde{o})] \\
&\approx \mathbb{E}_{Q(\pi)}[\ln Q(\pi) + F(\pi) - \ln P(\pi)] \\
P(\pi) &= Cat(\boldsymbol{\pi}_0) \\
Q(\pi) &= Cat(\boldsymbol{\pi}) \\
\boldsymbol{\pi}_0 &= \sigma(\ln \mathbf{E} - \mathbf{G})
\end{aligned} \qquad (B.7)$$

The approximate equality in the second line comes from equation B.2. Here, $\mathbf{E}$ is a vector of fixed beliefs about policies (this may be thought of as a *bias*, or *habit*, term), while $\mathbf{G}$ is the expected free energy for each policy. As before, we can now write the free energy in terms of sufficient statistics:

$$F = \boldsymbol{\pi} \cdot (\ln \boldsymbol{\pi} - \ln \mathbf{E} + \mathbf{F} + \mathbf{G}) \qquad (B.8)$$

Here, $\mathbf{F}$ is a vector whose elements are $\mathbf{F}_\pi$ as defined in equation B.4. Taking the gradients, we find the optimal update for beliefs about policies (i.e., planning):

$$\nabla_\pi F = 0 \Leftrightarrow$$
$$\pi = \sigma(\ln \mathbf{E} - \mathbf{F} - \mathbf{G}) \tag{B.9}$$

B.2.3 Learning

To enable learning, we need to incorporate prior beliefs about the parameters of the probability distributions that comprise the generative model. As these are expressed as categorical distributions, the appropriate (conjugate) choice of prior is a Dirichlet distribution. Taking the prior over initial states as an example, the terms in the free energy that depend on the expected (log) prior include the following:

$$F = \cdots + D_{KL}[Q(D)\,\|\,P(D)] - \mathbb{E}_{Q(s_1)Q(D)}[\ln P(s_1|D)]$$
$$= \cdots + (\mathbf{d} - d) \cdot \mathbb{E}_{Q(D)}[\ln \mathbf{D}] - \mathbf{s}_1 \cdot \mathbb{E}_{Q(D)}[\ln \mathbf{D}]$$

$$\mathbb{E}_{Q(D)}[\ln \mathbf{D}] = \psi(\mathbf{d}) - \psi(\mathbf{d}_0) \tag{B.10}$$
$$\mathbf{d}_0 = \sum_i \mathbf{d}_i$$
$$Q(D) \triangleq Dir(\mathbf{d})$$
$$P(D) \triangleq Dir(d)$$

Equation B.10 highlights in the third equality a useful identity. The expectation of the log of a Dirichlet distributed variable is the difference between two digamma functions (ψ)—where the digamma function is the derivative of a gamma function. We can use equation B.10 to find the free energy minimum:

$$\nabla_{\mathbb{E}[\ln \mathbf{D}]}F = \mathbf{d} - d - \mathbf{s}_1 = 0 \Leftrightarrow \mathbf{d} = d + \mathbf{s}_1 \tag{B.11}$$

This gives a simple scheme that may be used to update prior Dirichlet parameters to their posterior values. Very similar update rules apply for the other probability distributions that comprise the generative model:

$$\mathbf{a} = a + \sum_\tau o_\tau \otimes \mathbf{s}_\tau$$
$$\mathbf{b}_{\pi\tau} = b_{\pi\tau} + \sum_\tau \mathbf{s}_{\pi\tau} \otimes \mathbf{s}_{\pi\tau-1}$$
$$\mathbf{c} = c + \sum_\tau o_\tau \tag{B.12}$$
$$\mathbf{d} = d + \mathbf{s}_1$$
$$\mathbf{e} = e + \pi$$

These simply say that when the thing predicted by the relevant term in the probability distribution comes to pass (which may be a combination of two things for conditional probabilities), we simply augment that element of the probability array to signal that it is more likely to happen again in the future.

B.2.4 Precision

In some settings, it may be convenient to parameterize the generative model in a slightly different way. One option here is to use a Gibbs measure, where probability distributions are equipped with an inverse temperature parameter that plays the role of a precision. Most commonly, this is done for the precision (γ) over policies:

$$
\begin{aligned}
P(\pi \,|\, \gamma) &= Cat(\pi_0) \\
\pi_0 &= \sigma(-\gamma \mathbf{G})
\end{aligned}
\tag{B.13}
$$

For simplicity, we omit the $\mathbf{E}$ vector for this section. In what follows, we will also consider a precision for the likelihood (ζ) and for transitions (ω). The prior distribution over precision parameters is assumed to be a gamma distribution:

$$
\begin{aligned}
P(\zeta) &\propto \beta_\zeta \exp\!\left(-\beta_\zeta \zeta\right) \\
P(\omega) &\propto \beta_\omega \exp\!\left(-\beta_\omega \omega\right) \\
P(\gamma) &\propto \beta_\gamma \exp\!\left(-\beta_\gamma \gamma\right)
\end{aligned}
\tag{B.14}
$$

The approximate posterior distributions have the same (gamma distribution) form, and we will use a bold beta hyper-parameter to distinguish between the sufficient statistics of the posterior and prior above. A useful property of the gamma distribution, when parameterized in this way, is the following:

$$
\begin{aligned}
\zeta &= \mathbb{E}_{Q(\zeta)}[\zeta] = \boldsymbol{\beta}_\zeta^{-1} \\
\omega &= \mathbb{E}_{Q(\omega)}[\omega] = \boldsymbol{\beta}_\omega^{-1} \\
\gamma &= \mathbb{E}_{Q(\gamma)}[\gamma] = \boldsymbol{\beta}_\gamma^{-1}
\end{aligned}
\tag{B.15}
$$

Having defined these distributions, we can write the variational free energy:

$$
\begin{aligned}
F = \mathbb{E}_Q[F(\pi, \zeta, \omega) &+ D_{KL}[Q(\pi) \,\|\, P(\pi \,|\, \gamma)]] \\
&+ D_{KL}[Q(\gamma) \,\|\, P(\gamma)] + D_{KL}[Q(\omega) \,\|\, P(\omega)] + D_{KL}[Q(\zeta) \,\|\, P(\zeta)]
\end{aligned}
\tag{B.16}
$$

This can be expressed in terms of its sufficient statistics (omitting constants):

$$F = \pi \cdot (F + \ln \pi + \gamma \cdot G + \ln Z(\gamma)) + \ln \beta_\gamma + \ln \beta_\omega + \ln \beta_\zeta$$
$$- \ln \beta_\gamma - \ln \beta_\omega - \ln \beta_\zeta + \gamma \beta_\gamma + \omega \beta_\omega + \zeta \beta_\zeta \tag{B.17}$$
$$F_\pi \approx -\sum_\tau s_{\pi\tau} \cdot \left(\zeta \ln A \cdot o_\tau + \omega \ln B_{\pi\tau} s_{\pi\tau-1} - \ln Z(\zeta) \cdot o_\tau - \ln Z(\omega) s_{\pi\tau-1} \right)$$

In equation B.17, Z represent partition functions (i.e., normalizing constants) given by the following:

$$Z(\zeta)_j = \sum_i (A_{ij})^\zeta$$
$$Z(\omega)_j = \sum_i (B_{\pi\tau ij})^\omega$$
$$Z(\gamma) = \sum_\pi \exp(-\gamma \cdot G_\pi)$$
$$\Rightarrow$$
$$\partial_\zeta \ln Z(\zeta) s_\tau = o_\tau^\zeta \cdot \ln A \tag{B.18}$$
$$\partial_\omega \ln Z(\omega) s_{\pi\tau-1} = s_{\pi\tau}^\omega \cdot \ln B_\pi$$
$$\partial_\gamma \ln Z(\gamma) = -\pi_0 \cdot G$$
$$o_\tau^\zeta \triangleq \sigma(\zeta \ln A) s_\tau$$
$$s_{\pi\tau}^\omega \triangleq \sigma(\omega \ln B_{\pi\tau}) s_{\pi\tau-1}$$
$$\pi_0 \triangleq \sigma(-\gamma G)$$

Taking the partial derivative[2] with respect to the expected precisions gives this:

$$\left\{ \begin{matrix} \partial_\zeta F \\ \partial_\omega F \\ \partial_\gamma F \end{matrix} \right\} = 0 \Leftrightarrow \left\{ \begin{matrix} \beta_\zeta \\ \beta_\omega \\ \beta_\gamma \end{matrix} \right\} = \left\{ \begin{matrix} \sum_\tau (o_\tau^\zeta - o_\tau) \cdot \ln A + \beta_\zeta \\ \sum_\tau \pi \cdot (s_{\pi\tau}^\omega - s_{\pi\tau}) \cdot \ln B_\pi s_{\pi\tau-1} + \beta_\omega \\ (\pi - \pi_0) \cdot G + \beta_\gamma \end{matrix} \right\} \tag{B.19}$$

Expressing these updates as biologically plausible gradient descents gives the resulting equations:

$$\left\{ \begin{matrix} \dot{\beta}_\zeta \\ \dot{\beta}_\omega \\ \dot{\beta}_\gamma \end{matrix} \right\} = \left\{ \begin{matrix} \sum_\tau (o_\tau^\zeta - o_\tau) \cdot \ln A + \beta_\zeta - \beta_\zeta \\ \sum_\tau \pi \cdot (s_{\pi\tau}^\omega - s_{\pi\tau}) \cdot \ln B_\pi s_{\pi\tau-1} + \beta_\omega - \beta_\omega \\ (\pi - \pi_0) \cdot G + \beta_\gamma - \beta_\gamma \end{matrix} \right\} \tag{B.20}$$

Note that the dimensionality implies a (row) vector of precisions for A, where each state (column of A) is associated with its own precision parameter.

B.2.5   Expected Free Energy

Expected free energy is discussed extensively in the main text of the book. In this section, we supplement this discussion with two things. First, we offer a brief outline of current thinking as to why this is the appropriate quantity to define prior beliefs about policies. Second, we touch on the implementational details for computing this quantity.

While numerical simulations (of the sort illustrated in chapter 7) have established that expected free energy is useful, the question of *why* it is useful is still an active research area. Anticipating that this discussion will continue to evolve, here we set down a brief summary of the most parsimonious explanation at the time of writing (Da Costa et al. 2020; Friston, Da Costa et al. 2020). The starting point is to stipulate that a system attains some steady state (see section A.5.2) or, equivalently, fulfills its preferences (defined here in relation to latent states) at some future time (τ):

$$Q(s_\tau) = \mathbb{E}_{Q(\pi)}\big[Q(s_\tau \,|\, \pi)\big] = P(s_\tau \,|\, C) \tag{B.21}$$

Our challenge is to find the $Q(\pi)$ that satisfies equation B.21. To do so, we note that equation B.21 implies the following:

$$D_{KL}\big[Q(\pi \,|\, s_\tau)Q(s_\tau) \,\|\, Q(\pi \,|\, s_\tau)P(s_\tau \,|\, C)\big] = 0$$
$$\Rightarrow \tag{B.22}$$
$$\mathbb{E}_{Q(\pi, s_\tau)}\big[\ln Q(\pi, s_\tau)\big] = \mathbb{E}_{Q(\pi, s_\tau)}\big[\ln Q(\pi \,|\, s_\tau) + \ln P(s_\tau \,|\, C)\big]$$

We next factorize the left-hand side so as to isolate the $Q(\pi)$ term we are interested in:

$$\mathbb{E}_{Q(\pi)}\big[\ln Q(\pi)\big] = \mathbb{E}_{Q(\pi, s_\tau)}\big[\ln Q(\pi \,|\, s_\tau) + \ln P(s_\tau \,|\, C) - \ln Q(s_\tau \,|\, \pi)\big] \tag{B.23}$$

We define a variable α that represents the ratio of two entropies:

$$\alpha = \frac{\mathbb{E}_{Q(s_\tau)}\big[H[Q(\pi \,|\, s_\tau)]\big]}{\mathbb{E}_{Q(s_\tau, \pi)}\big[H[P(o_\tau \,|\, s_\tau)]\big]} \tag{B.24}$$

Heuristically, α expresses the relative range of behavioral outputs (i.e., policies) that are plausible in a given state, compared to the range of outcomes expected in that same state. If very large, this might describe a creature whose behavior bears little relationship to the state of their world, despite highly precise sensory observations being generated by that world. When very small, this might describe a creature who always behaves the same way when it knows the state of the world but is rarely offered precise data about that world. Here, we will stipulate that we are interested in systems whose $\alpha = 1$—implying a relatively balanced and symmetrical exchange with their world. This renders the two entropies in equation B.24 equal. Returning to equation B.23:

$$
\begin{aligned}
\mathbb{E}_{Q(\pi)}[\ln Q(\pi)] &= -\mathbb{E}_{Q(s_\tau)}\big[H[Q(\pi|s_\tau)]\big] \\
&\quad + \mathbb{E}_{Q(\pi,s_\tau)}[\ln P(s_\tau|C) - \ln Q(s_\tau|\pi)] \\
&= -\mathbb{E}_{Q(s_\tau,\pi)}\big[H[P(o_\tau|s_\tau)]\big] \\
&\quad - \mathbb{E}_{Q(\pi)}\big[D_{KL}[Q(s_\tau|\pi)\,\|\,P(s_\tau|C)]\big]
\end{aligned}
\tag{B.25}
$$

The second line follows from the first and from equation B.24 with $\alpha = 1$. We now see that equation B.25, and therefore equation B.21, is satisfied by choosing the following:

$$
\ln Q(\pi) = -\underbrace{\mathbb{E}_{Q(s_\tau|\pi)}\big[H[P(o_\tau|s_\tau)]\big]}_{\text{Expected ambiguity}} - \underbrace{D_{KL}[Q(s_\tau|\pi)\,\|\,P(s_\tau|C)]}_{\text{Risk}}
\tag{B.26}
$$

Our final step is to note the relationship between the quantity on the right-hand side and the expected free energy—with preferences defined in terms of observations in place of states:

$$
\begin{aligned}
&\mathbb{E}_{Q(s_\tau|\pi)}\big[H[P(o_\tau|s_\tau)]\big] + D_{KL}[Q(s_\tau|\pi)\,\|\,P(s_\tau|C)] \\
&= \mathbb{E}_{Q(s_\tau|\pi)}\big[H[P(o_\tau|s_\tau)]\big] + D_{KL}[Q(s_\tau|\pi)\,\|\,P(s_\tau|C)] \\
&\quad + \underbrace{\mathbb{E}_{Q(s_\tau|\pi)P(o_\tau|s_\tau)}[\ln P(o_\tau|s_\tau)] - \mathbb{E}_{Q(s_\tau|\pi)P(o_\tau|s_\tau)}[\ln P(o_\tau|s_\tau)]}_{=0} \\
&= \mathbb{E}_{Q(s_\tau|\pi)}\big[H[P(o_\tau|s_\tau)]\big] + D_{KL}[Q(o_\tau,s_\tau|\pi)\,\|\,P(o_\tau,s_\tau|C)] \\
&= \mathbb{E}_{Q(s_\tau|\pi)}\big[H[P(o_\tau|s_\tau)]\big] + D_{KL}[Q(o_\tau|\pi)\,\|\,P(o_\tau|C)] \\
&\quad + \mathbb{E}_{Q(o_\tau|\pi)}\big[D_{KL}[Q(s_\tau|o_\tau,\pi)\,\|\,P(s_\tau|o_\tau,C)]\big] \\
&\geq \mathbb{E}_{Q(s_\tau|\pi)}\big[H[P(o_\tau|s_\tau)]\big] + D_{KL}[Q(o_\tau|\pi)\,\|\,P(o_\tau|C)] = G(\pi)
\end{aligned}
\tag{B.27}
$$

The steps in equation B.27 have (perhaps a little tediously) been included in some detail as our experience is that people often struggle with this result. The inequality in the final line arises from the omission of the (nonnegative) KL-Divergence in the previous line. The key result here is that the ambiguity and risk minimized for the most plausible policies in equation B.26 acts as an upper bound on the expected free energy used throughout this book.

We now turn to the question of computational implementation. Here we can simply appeal to the linear algebraic identities we saw in section A.2. It is straightforward, if we express preferences as a vector of prior probabilities (**C**), to express the pragmatic term of the expected free energy as follows:

$$\mathbb{E}_{Q(o_\tau|\pi)}\left[\ln P(o_\tau|C)\right] = \mathbf{o}_{\pi\tau}\cdot\ln\mathbf{C}_\tau$$
$$Q(o_\tau|\pi) = Cat(\mathbf{o}_{\pi\tau}) \tag{B.28}$$
$$\mathbf{o}_{\pi\tau} = \mathbf{A}\mathbf{s}_{\pi\tau}$$

The expected information gain associated with hidden states (i.e., salience, epistemic value, or Bayesian surprise) is expressed in terms of the difference between two entropies:

$$H[Q(o_\tau|\pi)] - \mathbb{E}_{Q(s_\tau|\pi)}[H[P(o_\tau|s_\tau)]]$$
$$= -\mathbf{o}_\pi\cdot\ln\mathbf{o}_\pi - \mathbf{H}\cdot\mathbf{s}_{\pi\tau} \tag{B.29}$$
$$\mathbf{H}\triangleq -diag(\mathbf{A}\cdot\ln\mathbf{A})$$

See Section A.2.3 for an explanation of the last line. Putting equations B.28 and B.29 together, the expected free energy is this:

$$G_\pi = \sum_\tau G_{\pi\tau}$$
$$G_{\pi\tau} = \mathbf{H}\cdot\mathbf{s}_{\pi\tau} + \mathbf{o}_{\pi\tau}\cdot\left(\ln\mathbf{o}_{\pi\tau} - \ln\mathbf{C}_\tau\right) \tag{B.30}$$

When we need to account for active learning, we supplement this with parameter information gain. The information gain associated with parameters of the generative model (i.e., novelty) may be derived as follows. Using the KL-Divergence between two Dirichlet distributions, we can express the information gain that would occur following a given state-outcome combination:

$$W_{ij} \triangleq D_{KL}[P(A_{ij}|o=i,s=j) \| P(A_{ij})]$$
$$= \underbrace{\left(\ln\Gamma(a_{ij}) - \ln\Gamma(a_{ij}+1)\right)}_{-\ln a_{ij}} + \underbrace{\left(\ln\Gamma(a_{0j}+1) - \ln\Gamma(a_{0j})\right)}_{+\ln a_{0j}} \tag{B.31}$$
$$+ \psi(a_{ij}+1) - \psi(a_{0j}+1)$$
$$a_{0j} \triangleq \sum_i a_{ij}$$

Here we have used the fact that if we knew a given state-outcome combination had occurred, we would add 1 to the associated Dirichlet parameter. This lets us use a standard identity of a log gamma function (as indicated by the underbraces) to simplify the expression:

$$W_{ij} = \ln\frac{a_{0j}}{a_{ij}} + \frac{\partial_{a_{ij}}\Gamma(a_{ij}+1)}{a_{ij}\Gamma(a_{ij})} - \frac{\partial_{a_{0j}}\Gamma(a_{0j}+1)}{a_{0j}\Gamma(a_{0j})}$$

$$= \ln\frac{a_{0j}}{a_{ij}} + \frac{1}{a_{ij}} + \frac{\partial_{a_{ij}}\Gamma(a_{ij})}{\Gamma(a_{ij})} - \frac{1}{a_{0j}} - \frac{\partial_{a_{0j}}\Gamma(a_{0j})}{\Gamma(a_{0j})} \tag{B.32}$$

Here we have used the identity $x\Gamma(x) = \Gamma(x+1)$ and an application of the product rule. We next use the identity $\psi(x)\Gamma(x) = \partial_x\Gamma(x)$ and the approximation $\psi(x) \approx \ln x - (2x)^{-1}$ to simplify this:

$$= \frac{1}{a_{ij}} - \frac{1}{a_{0j}} + \ln\frac{a_{0j}}{a_{ij}} + \psi(a_{ij}) - \psi(a_{0j})$$

$$\approx \frac{1}{2a_{ij}} - \frac{1}{2a_{0j}} \tag{B.33}$$

The expected information gain is then as follows:

$$\mathbb{E}_{Q(o_\tau,s_\tau|\pi)}[D_{KL}[P(A|o_\tau,s_\tau)\,\|\,P(A)]] \approx \mathbf{o}_{\pi\tau} \cdot \mathbf{Ws}_{\pi\tau} \tag{B.34}$$

This simple expression then augments the expected free energy to ensure novelty-seeking behavior in addition to pragmatic and salient choices.

B.2.6 Bayesian Model Reduction

In chapter 7, we briefly touch on the idea of structure learning and model reduction. We take the opportunity to unpack the principles in a little more depth here. Bayesian model reduction is a technique used to compare alternative models that differ only in their priors. Through Bayes' theorem (see chapter 2), we can express the ratio of the joint probability of data (y) and some parameters (θ) between two alternative models in two different ways:

$$\frac{P(y,\theta)}{\tilde{P}(y,\theta)} = \frac{P(y|\theta)P(\theta)}{\tilde{P}(y|\theta)\tilde{P}(\theta)} = \frac{P(\theta|y)P(y)}{\tilde{P}(\theta|y)\tilde{P}(y)} \tag{B.35}$$

If the only difference between the two models is the prior (i.e., $P(y|\theta) = \tilde{P}(y|\theta)$), then we can cancel the likelihood terms. On rearranging, this gives an expression for the posterior probability under alternative (reduced) priors in terms of the posterior probability under the original (full) priors:

$$\tilde{P}(\theta|y) = \frac{P(\theta|y)P(y)\tilde{P}(\theta)}{\tilde{P}(y)P(\theta)} \tag{B.36}$$

Integrating both sides with respect to the parameters gives this:

$$1 = \frac{P(y)}{\tilde{P}(y)} \mathbb{E}_{P(\theta|y)} \left[\frac{\tilde{P}(\theta)}{P(\theta)} \right] \Rightarrow$$

$$\ln \tilde{P}(y) = \ln P(y) + \ln \mathbb{E}_{P(\theta|y)} \left[\frac{\tilde{P}(\theta)}{P(\theta)} \right] \tag{B.37}$$

Substituting back into equation B.36 gives this:

$$\ln \tilde{P}(\theta|y) = \ln P(\theta|y) + \ln \tilde{P}(\theta) - \ln P(\theta) - \ln \mathbb{E}_{P(\theta|y)} \left[\frac{\tilde{P}(\theta)}{P(\theta)} \right] \tag{B.38}$$

Together, equations B.37 and B.38 mean that we can find the model evidence and posterior we would have got, had we used a given reduced prior, using the results from inverting a full model. We can reexpress these equations in terms of the variational quantities introduced in chapter 4:

$$F[P(\theta)] - F[\tilde{P}(\theta)] = \ln \mathbb{E}_{Q(\theta)} \left[\frac{\tilde{P}(\theta)}{P(\theta)} \right]$$

$$\ln \tilde{Q}(\theta) = \ln Q(\theta) + \ln \tilde{P}(\theta) - \ln P(\theta) - \ln \mathbb{E}_{Q(\theta)} \left[\frac{\tilde{P}(\theta)}{P(\theta)} \right] \tag{B.39}$$

For reference, we offer the form of equation B.39 under two different kinds of prior. The first is a normal distribution:[3]

$$P(\theta) = \mathcal{N}(\eta, \Sigma)$$

$$\tilde{P}(\theta) = \mathcal{N}(\tilde{\eta}, \tilde{\Sigma})$$

$$Q(\theta) = \mathcal{N}(\mu, C)$$

$$\tilde{Q}(\theta) = \mathcal{N}(\tilde{\mu}, \tilde{C}) \tag{B.40}$$

$$\tilde{C}^{-1} = \tilde{P} = P + \tilde{\Pi} - \Pi$$

$$\tilde{\mu} = \tilde{C}(P\mu + \tilde{\Pi}\tilde{\eta} - \Pi\eta)$$

$$\Delta F = -\tfrac{1}{2} \ln | \tilde{\Pi} P \tilde{C} \Sigma | + \tfrac{1}{2}(\mu \cdot P\mu + \tilde{\eta} \cdot \tilde{\Pi}\tilde{\eta} - \eta \cdot \Pi\eta - \tilde{\mu} \cdot \tilde{P}\tilde{\mu})$$

Practically, this is used in the setting of mixed models, with a continuous and a categorical component. If each categorical outcome of the latter is associated with a continuous prior, we can efficiently evaluate the evidence for each of these priors (and therefore categorical outcomes) without having to invert each model in turn. See chapter 8 for an example.

More often, in a purely POMDP setting, we may be interested in comparing hypotheses about Dirichlet prior distributions. This has been used to simulate pruning of elements in a probability matrix as a metaphor for synaptic pruning during sleep (Friston, Lin et al. 2017). The form of Bayesian model reduction for Dirichlet distributions is as follows:

$$P(\theta) = Dir(a)$$
$$\tilde{P}(\theta) = Dir(\tilde{a})$$
$$Q(\theta) = Dir(\mathbf{a})$$
$$\tilde{Q}(\theta) = Dir(\tilde{\mathbf{a}}) \tag{B.41}$$

$$\tilde{\mathbf{a}} = \mathbf{a} + \tilde{a} - a$$

$$\Delta F = \ln B(\mathbf{a}) - \ln B(\tilde{\mathbf{a}}) + \ln B(\tilde{a}) - \ln B(a)$$

In this expression, B denotes a beta function. Similar results can be derived for a range of distributions (see Friston, Parr, and Zeidman 2018), but normal and Dirichlet priors are the most commonly encountered in Active Inference.

B.3   (Active) Generalized Filtering

We now move from the categorical inferences under a POMDP model to the continuous domain. This is where some of the preliminaries from appendix A really start to pay off. We will exploit the Laplace approximation and generalized coordinates of motion, both presented in section A.3. In addition, we will need to construct precision matrices including different orders of generalized motion, as we saw in section A.5.2. From equations A.33 and A.34 we can write the free energy under the Laplace approximation as follows:

$$F[q, \tilde{y}] \approx -\frac{1}{2}\ln(2\pi)^k \left|\tilde{\Sigma}\right| - \ln p(\tilde{y}, \tilde{\mu})$$
$$q(\tilde{x}) = \mathcal{N}(\tilde{\mu}, \tilde{\Sigma}^{-1}) \tag{B.42}$$
$$\tilde{\Sigma}^{-1} = -\nabla_{\tilde{x}}\left(\nabla_{\tilde{x}} \ln p(\tilde{y}, \tilde{x})\right)^T\Big|_{\tilde{x}=\tilde{\mu}}$$

Here we have expressed the free energy, using the Laplace assumption, for a model defined in generalized coordinates. Under the Laplace assumption, the only term in the first line that varies with μ is the last one. This is the term expressing the generative model. Our next step is to specify the form of the generative model:

$$p(\tilde{y},\tilde{x},\tilde{v}) = p(\tilde{y} \mid \tilde{x},\tilde{v})p(\tilde{x} \mid \tilde{v})p(\tilde{v})$$

$$p(\tilde{y} \mid \tilde{x},\tilde{v}) = \mathcal{N}(\tilde{g}(\tilde{x},\tilde{v}),\tilde{\Pi}_y)$$

$$p(\tilde{x} \mid \tilde{v}) = \mathcal{N}(D \cdot \tilde{f}(\tilde{x},\tilde{v}),\tilde{\Pi}_x)$$

$$p(\tilde{v}) = \mathcal{N}(\tilde{\eta},\tilde{\Pi}_x) \qquad\qquad (B.43)$$

$$D\tilde{x} = \tilde{f}(\tilde{x},\tilde{v}) + \tilde{\omega}_x$$

$$\tilde{y} = \tilde{g}(\tilde{x},\tilde{v}) + \tilde{\omega}_y$$

Note we now have two hidden variables, x and v. The difference is that the former depends on an equation of motion (f), while the latter depends on a static prior. The D operator in the penultimate line is a matrix with ones above the leading diagonal. In generalized coordinates, this is equivalent to taking a temporal derivative, as each element in the vector of temporal derivatives is shifted up by one. The generalized precisions are constructed as in section A.5.3. Substituting the quantities of equation B.43 into B.42, we have the following:

$$F[q,\tilde{y}] = \underbrace{\tfrac{1}{2}\tilde{\varepsilon}_y \cdot \tilde{\Pi}_y \tilde{\varepsilon}_y}_{-\ln p(\tilde{y}\mid\tilde{\mu}_x,\tilde{\mu}_v)} + \underbrace{\tfrac{1}{2}\tilde{\varepsilon}_x \cdot \tilde{\Pi}_x \tilde{\varepsilon}_x}_{-\ln p(\tilde{\mu}_x\mid\tilde{\mu}_v)} + \underbrace{\tfrac{1}{2}\tilde{\varepsilon}_v \cdot \tilde{\Pi}_v \tilde{\varepsilon}_v}_{-\ln p(\tilde{\mu}_v)}$$

$$\tilde{\varepsilon}_y \triangleq \tilde{y} - \tilde{g}(\tilde{\mu}_x,\tilde{\mu}_v)$$

$$\tilde{\varepsilon}_x \triangleq D\tilde{\mu}_x - \tilde{f}(\tilde{\mu}_x,\tilde{\mu}_v) \qquad\qquad (B.44)$$

$$\tilde{\varepsilon}_v \triangleq \tilde{\mu}_v - \tilde{\eta}$$

We have omitted all constants with respect to μ. From equation B.44, we can find the gradients of the free energy (using identities introduced in section A.2.2):

$$\nabla_{\tilde{\mu}_x} F[q,\tilde{y}] = -\nabla_{\tilde{\mu}_x}\tilde{g} \cdot \tilde{\Pi}_y \tilde{\varepsilon}_y + D \cdot \tilde{\Pi}_x \tilde{\varepsilon}_x - \nabla_{\tilde{\mu}_x}\tilde{f} \cdot \tilde{\Pi}_x \tilde{\varepsilon}_x$$

$$\nabla_{\tilde{\mu}_v} F[q,\tilde{y}] = -\nabla_{\tilde{\mu}_v}\tilde{g} \cdot \tilde{\Pi}_y \tilde{\varepsilon}_y - \nabla_{\tilde{\mu}_v}\tilde{f} \cdot \tilde{\Pi}_x \tilde{\varepsilon}_x + \tilde{\Pi}_v \tilde{\varepsilon}_v \qquad (B.45)$$

We could now specify a gradient descent to find the values of μ that minimize free energy. However, this would imply that when the free energy is minimized, μ becomes static. Clearly this is suboptimal if we believe higher orders of motion to be nonzero. To account for this, we can express a gradient descent in a moving frame of reference, such that when the free energy is minimized, μ continues to move with velocity μ':

$$\dot{\tilde{\mu}}_x - D\tilde{\mu}_x = \nabla_{\tilde{\mu}_x}\tilde{g} \cdot \tilde{\Pi}_y \tilde{\varepsilon}_y - D \cdot \tilde{\Pi}_x \tilde{\varepsilon}_x + \nabla_{\tilde{\mu}_x}\tilde{f} \cdot \tilde{\Pi}_x \tilde{\varepsilon}_x$$
$$\dot{\tilde{\mu}}_v - D\tilde{\mu}_v = \nabla_{\tilde{\mu}_v}\tilde{g} \cdot \tilde{\Pi}_y \tilde{\varepsilon}_y + \nabla_{\tilde{\mu}_v}\tilde{f} \cdot \tilde{\Pi}_x \tilde{\varepsilon}_x - \tilde{\Pi}_v \tilde{\varepsilon}_v$$

$$(B.46)$$

Equation B.46 specifies a predictive coding scheme, in which prediction errors drive updates in expectations, resolving those errors. These schemes may be extended to include multiple hierarchical levels by duplicating the equations of B.46 for an additional level but replacing y with v from the lower level:

$$\dot{\tilde{\mu}}_x^{(i)} - D\tilde{\mu}_x^{(i)} = \nabla_{\tilde{\mu}_x^{(i)}}\tilde{g} \cdot \tilde{\Pi}_v^{(i-1)} \tilde{\varepsilon}_v^{(i-1)} - D \cdot \tilde{\Pi}_x^{(i)} \tilde{\varepsilon}_x^{(i)} + \nabla_{\tilde{\mu}_x^{(i)}}\tilde{f}^{(i)} \cdot \tilde{\Pi}_x^{(i)} \tilde{\varepsilon}_x^{(i)}$$
$$\dot{\tilde{\mu}}_v^{(i)} - D\tilde{\mu}_v^{(i)} = \nabla_{\tilde{\mu}_v^{(i)}}\tilde{g}^{(i)} \cdot \tilde{\Pi}_v^{(i-1)} \tilde{\varepsilon}_v^{(i-1)} + \nabla_{\tilde{\mu}_v^{(i)}}\tilde{f}^{(i)} \cdot \tilde{\Pi}_x^{(i)} \tilde{\varepsilon}_x^{(i)} + \tilde{\Pi}_v^{(i)} \tilde{\varepsilon}_v^{(i)}$$
$$\tilde{\varepsilon}_v^{(i)} \triangleq \tilde{\mu}_v^{(i)} - \tilde{g}^{(i+1)}(\tilde{\mu}_x^{(i+1)}, \tilde{\mu}_v^{(i+1)})$$
$$\tilde{\varepsilon}_x^{(i)} \triangleq D\tilde{\mu}_x^{(i)} - \tilde{f}^{(i)}(\tilde{\mu}_x^{(i)}, \tilde{\mu}_v^{(i)})$$

$$(B.47)$$

Under Active Inference, the free energy is minimized by perception but also by action. As the only thing action changes is the sensory input (y), most of the terms in equation B.44 are irrelevant for action. Minimizing the free energy with respect to action gives this:

$$\dot{u} = -\nabla_u \tilde{y}(u) \cdot \tilde{\Pi}_y \tilde{\varepsilon}_y$$

$$(B.48)$$

Together, equations B.47 and B.48 provide a very general description of Active Inference for continuous state-space models. We will not discuss the issue of learning or mixed models in this section, as these are summarized in boxes 8.2 and 8.3, respectively.

Appendix C: An Annotated Example of the Matlab Code

C.1   Introduction

Here we provide an annotated example of the Matlab code—using the standard inversion and plotting routines available in SPM12—required to specify and solve a generative model. This reproduces the T-maze foraging example in chapter 7. This appendix, a little dry on its own, will be most useful for readers who attempt to implement this code in Matlab so that they can see it working. We recommend trying to "break" this demo by playing with different parameter values and changing the generative model. Only by doing this will an intuitive sense of the mechanics of Active Inference develop.

C.2   Preliminaries

We assume that readers have some familiarity with Matlab and have successfully downloaded the SPM12 software package from https://www.fil.ion .ucl.ac.uk/spm/. The first step is to ensure that the folder containing the SPM12 functions is added to the Matlab path. We then open a Matlab script and begin writing our demo by defining a function and giving it a name (here, demo_AI_book):

```
function demo_AI_book
rng default
```

Figure C.1

The second line in figure C.1 sets the random number generator to a default initial seed so that the same random numbers are generated each time the function is used. We normally do this for demos, as this ensures reproducibility. However, this can be omitted if we are instead running this function multiple times to compute some summary statistics of behavior over multiple trials. It is good practice to include some comments here that tell people about the script. Given that this appendix is devoted to annotation of this script, we omit this documentation here.

Next, we define some of the important constants we will use later. The advantage of listing these together here is that it is easy for us to find them in case we want to perturb them later. We define two parameters that will play the role of probabilities, whose role will become clear in section C.3 (figure C.2):

```
a       = .98;
b       = 1-a;
```

Figure C.2

This definition ensures $a + b = 1$. Now we are ready to set up the **A**, **B**, **C**, and **D** matrices and vectors that define a generative model. In this simple simulation, we assume these parameters are the same in the generative model and the generative process.

C.3   The Likelihood

Our focus here is on how to formalize a likelihood matrix in Matlab, so we will not devote a great deal of space to describing the generative model or the paradigm it describes. (This description is given in section 7.2.) Our aim is to translate the likelihood matrices of figures 7.4 and 7.5 into a language that our inversion routines will understand. We start with A^1, which is written as A{1}, where the term inside curly { } brackets corresponds to the superscript (i.e., outcome modality); see figure C.3. The elements of the matrix (or tensor, more technically) are addressable using three indexes. These are the outcome, the first (location) hidden state, and the second (context) hidden state. These appear inside normal ( ) brackets. The matrix for the first level of the second hidden state factor—the context in which the attractive stimulus is on the right—is then specified with a 1 index in

the third position. The context in which it is on the left is specified with a
2 index in this position:

```
A{1}  (:,:,1) = [...
         1 0 0 0;      % start
         0 0 0 0;      % right cue
         0 1 0 0;      % left cue
         0 0 1 0       % left
         0 0 0 1];     % right
A{1}  (:,:,2) = [...
         1 0 0 0;      % start
         0 1 0 0;      % right cue
         0 0 0 0;      % left cue
         0 0 1 0       % left
         0 0 0 1];     % right
```

Figure C.3

The rows are the outcomes, and the columns are the alternative levels of
the first hidden state factor. Comparison with the matrices displayed in
figures 7.4 and 7.5 should help clarify this syntax. The A^2 matrices are similarly defined for context. Here, we make us of the a and b we defined at the
top of the script (figure C.4):

```
A{2}  (:,:,1) = [...
         1 1 0 0;      % reward neutral
         0 0 a b;      % reward positive
         0 0 b a];     % reward negative
A{2}  (:,:,2) = [...
         1 1 0 0;      % reward neutral
         0 0 b a;      % reward positive
         0 0 a b];     % reward negative
```

Figure C.4

This completes our specification of **A.** We have probabilities defined for
every combination of outcome (rows) in two modalities (superscript or
curly bracket) for each combination of values for two hidden state factors
(second and third indices).

C.4   Transition Probabilities

Following **A**, we now specify **B**. As discussed in chapter 7, the superscript associated with the **B** matrix refers to the hidden state factor, as opposed to the outcome modalities indicated by the superscript for **A**; again, we use curly brackets as equivalent to the superscript (figure C.5). Recall the two factors are location (1) and context (2). Each matrix maps from the state at the previous time (column) to the current time (row). The **B** matrices can vary with each action if the state factor is controllable. This means that control states require an additional index specifying which action is taken:

```
B{1}(:,:,1)   = [1 1 0 0; 0 0 0 0;0 0 1 0;0 0 0 1];
B{1}(:,:,2)   = [0 0 0 0; 1 1 0 0;0 0 1 0;0 0 0 1];
B{1}(:,:,3)   = [0 0 0 0; 0 0 0 0;1 1 1 0;0 0 0 1];
B{1}(:,:,4)   = [0 0 0 0; 0 0 0 0;0 0 1 0;1 1 0 1];

B{2}    = eye(2);
```

Figure C.5

Here we have specified the controllable location state with each of the four actions available. Recall that there are four available actions, each determining a transition to one of the four locations. The right and left arms are absorbing states, meaning the rat must stay in these once entered. The semicolons here indicate the end of each row of the matrix. The contextual states are not under the creature's control, so we do not need to specify a different set of transition probabilities for each action. We simply define a single identity matrix, denoted by `eye` in Matlab. This implements the specification in figure 7.6 and equation 7.5.

C.5   Prior Preferences and Initial States

Following **B** is **C**. Here we return to a similar syntax as **A**: both superscripts and curly brackets pertain to the outcome modalities. In addition, we specify **D** as we did **B**: superscripts and curly brackets relate to hidden state factors (figure C.6). The number of rows of **C** and **D** must correspond to the number of rows in the associated **A** and **B** matrices, respectively.

```
C{1}   = [-1 -1 -1;
            0   0   0;
            0   0   0;
            0   0   0;
            0   0   0];
c      = 6;
C{2}   = [ 0   0   0;
            c   c   c;
           -c  -c  -c];
D{1}   = [1 0 0 0]';
D{2}   = [1 1]'/2;
```

Figure C.6

The preferences are specified in the **C** matrices in terms of log probabilities (which do not need to be normalized). A difference of six between two outcomes means the more probable outcome is exp(6) times more likely. Each row pertains to a different outcome. Columns correspond to different time steps. This allows for the possibility of time-varying preferences. If only one column is specified, it will be assumed that preferences are the same at each time. The **D** vectors specified here simply ascribe a probability to each state for the start of the trial. See equations 7.6 and 7.7 for comparison. Note that in Matlab, a transpose is denoted by ʹ.

C.6   The Policy Space

We specified the allowable actions implicitly through the **B** matrices. We could assume that the policies and actions are one and the same and that we select new policies every time step. Alternatively, we can specify policies as sequences of actions. One way of thinking about this is that each policy tells us which **B** matrix (indicated by the third index) is in play at each time step. We do this through specifying an array, V (figure C.7):

```
V(:,:,1) = [1 1 1 1 3 4 2 2 2 2;
            1 2 3 4 3 4 1 2 3 4];
V(:,:,2) =  1;
```

Figure C.7

The first index of V (i.e., the rows) represents the position in the action sequence. This means the first row is the first action and the second row is the second action. As actions cause transitions, a three-step model only requires two actions. The second index (i.e., the columns) represents the alternative policies that could be chosen. Finally, the third index is the hidden state factor. To aid intuition, $V(2,5,1) = 3$ means that the fifth policy option involves selecting the location $\mathbf{B}^1$ matrix associated with the third action to transition from the second to the third time step. While in simple simulations one may include all possible policies, in other cases one may select a subset. Note, however, that the selection of the available policies is a design choice in itself and has implications for the resulting behavior.

C.7   Putting It Together

Having specified our POMDP, we now bring it all together in a single mdp variable (figure C.8):

```
mdp.V = V;            % allowable policies
mdp.A = A;            % observation model
mdp.B = B;            % transition probabilities
mdp.C = C;            % preferred outcomes
mdp.D = D;            % prior over initial states
mdp.S = [1 1]';       % true initial state
```

Figure C.8

The final line here lets us specify the true hidden states that we wish to start with. This information is not available to our simulated creature, who must infer states on the basis of the outcomes generated by these states and its generative model (specified in the first five lines).

C.8   Simulation and Plotting

The heavy lifting is completed behind the scenes by the spm_MDP_VB_X function, which implements the message passing and policy selection described in chapters 4 and 7. In addition, it simulates the world our creature must contend with, including transitions between states and the generation of outcomes. We could have included many other options; for details, we refer readers to the documentation for this function in the Matlab script.

Once we have simulated a trial, we often want to find some graphical representation of the results. Plots of the sort shown in figures 7.2 and 7.7 can be automatically generated with standard plotting routines:

```
MDP = spm_MDP_VB_X(mdp);
spm_figure('GetWin','Figure 1'); clf
spm_MDP_VB_trial(MDP);

spm_figure('GetWin','Figure 2'); clf
spm_MDP_VB_LFP(MDP,[],1);

spm_figure('GetWin','Figure 3'); clf
spm_MDP_VB_LFP(MDP,[],2);
```

Figure C.9

The lines of code in figure C.9 will simulate and plot the results in three figures. The first figure provides an overall summary of the simulation, including states, outcomes, policies selected, and retrospective inferences. The second figure shows the electrophysiological correlates of belief updating for the first hidden state factor; the third figure does the same for the second factor. We have reproduced the outputs of the first lines of code in graphical form (figure C.10) to reassure you that everything is working. We could have reproduced the other two; however, we wanted to give you the opportunity to engage in Active Inference and reduce your uncertainty about what the plots show by running the script yourself.

This concludes the simple annotated example, which we hope sets you on the path to exploring the range of generative models that can be specified using the same principles and syntax. You can find further examples by typing "DEM" into the Matlab command line and selecting demos from the resulting graphical user interface. The demo script we have provided is a simplified version of the routine used in Friston, FitzGerald et al. (2017). For further details, including learning of the generative model over multiple trials, we refer you to that paper and to the DEM_demo_MDP_X.m script available in SPM12.

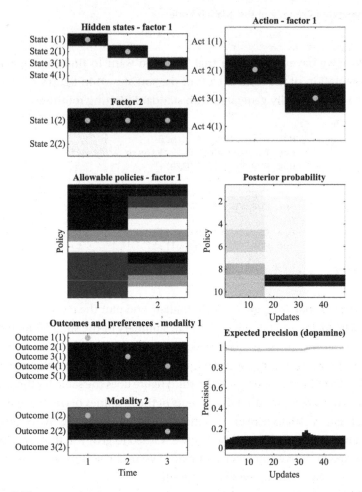

Figure C.10

Output obtained from the `spm_MDP_VB_trial` routine. *Upper left:* Beliefs held about each of the hidden state factors retrospectively (i.e., after all observations have been made). Black shading indicates a probability of one; white, of zero. Gray dots (which appear cyan when plotted in color) indicate the true states generated by the simulated environment; we see that the simulated rat accurately and confidently infers its location (Factor 1) and the context (Factor 2). *Middle left:* "Allowable policies" of the v variable specified in section C.6, showing a policy in each row: the first column is the first action taken; the second column, the second action. The different shades in each element represent different actions that could be chosen. *Lower left:* True outcomes (gray dots) in each modality. The background shading shows the C-matrices for each modality, with darker shades indicating preferred outcomes. *Upper right:* Inferred and selected actions. *Middle right:* Beliefs about each policy for each time step. *Bottom right:* Inferred precision of beliefs about the policies (gray line) with the rate of change plotted as a bar plot, reminiscent of the raster plots used to illustrate dopaminergic neuron firing in electrophysiology. Note that the time is specified in terms of updates; these refer to the iterations of a gradient descent on free energy. By default, there are sixteen updates per time step.

Notes

Chapter 1

1. The term *normative* means there is some evaluative standard against which behavior can be scored. Active Inference is normative in the sense that perception and action are scored by free energy—a quantity we will unpack throughout this and the next chapters.

2. *Bayes optimality* refers to a set of related concepts that deal with aspects of Bayes' theorem—something we will unpack in chapter 2. Broadly, it refers to any action that minimizes (or maximizes) the expected value of some cost (or utility) function given some observation. This encompasses Bayes-optimal experimental design, wherein an experiment (action) is chosen to maximize expected information gain.

Chapter 2

1. Like bits, nats are units of information. The choice of unit depends on whether we use a logarithm to the base 2 (bits) or a natural logarithm (nats).

2. *Support* is a technical term referring to the possible arguments for a distribution. For example, the support of a categorical probability distribution is a series of alternative states (i.e., event space) whose probability may be quantified. The support of a univariate normal distribution is the entire real number line.

3. The details of this table are not important for understanding Active Inference conceptually, but for interested readers, we briefly unpack the key points. The Support column tells us the set of variables whose surprise can be quantified using each distribution. This is the set of real numbers for the Gaussian distribution. For the multinomial distribution, the support comprises a group of K variables, each taking an integer value up to a maximum N, under the constraint that all elements in that group sum to N. For the Dirichlet distribution, the support includes any group of K real numbers between 0 and 1, where all elements in the group sum to 1. The gamma distribution quantifies the surprise of nonnegative real numbers. The Surprise column shows the

way in which the surprise can be calculated. This depends on constants (in addition to the random variable x) that control the shape of the underlying distribution.

4. Interestingly, resource limitations are not the only barrier to exact Bayesian inference. In the presence of complex models, exact inference may be analytically intractable, such that no additional resources could help solve the exact problem.

5. Like the KL-Divergence, entropy is a quantity from information theory. It is a measure of the dispersion (or uncertainty) of a probability distribution. Technically, it is the average of the negative log probability or average surprise.

6. *Complexity* as used here scores the degree to which we must depart from our prior beliefs about the world in order to explain data.

7. This is referred to as *accuracy* because an explanation's accuracy increases when a high log probability of outcomes, expected under the inferred hidden states, is assigned to observed data—i.e., when the predicted distribution of outcomes accurately captures the measured distribution.

Chapter 3

1. *Nonequilibrium* here refers to the absence of *detailed balance*. Detailed balance is the invariance of a system under time reversal once it has reached steady state. We can see that the system on the left of figure 3.3 does not possess detailed balance, as the trajectory tends to curve counterclockwise around the contours of surprise. If we were to play this back in reverse, the system would appear to rotate clockwise.

2. This is not the same as saying that surprise-minimizing systems must minimize their entropy. As we see in figure 3.3, the system does not tend toward an infinitely precise (point) distribution that would minimize entropy, but it maintains a consistent dispersion over time—bounding entropy from above and below.

3. The capital A is used to distinguish Action as a path integral of a Lagrangian from action as the dynamics of active states of a Markov blanket.

4. A Lagrangian is a function of a position and velocity that gives the difference between kinetic and potential energies. A Hamiltonian is related to (via a Legendre transform) and expresses the total energy of the system in terms of position and momentum.

Chapter 4

1. Here and throughout the chapter, the conditioning on the model is left implicit; hence, the model evidence is written as $P(y)$ and not $P(y|m)$.

2. Technically, this is true for any concave function, but we are concerned only with logarithms here.

3. An expectation is a weighted sum or integral of the term inside the square brackets; each term is weighted by the probability indicated by the subscript (see box 2.2).

4. In this book, we follow the physicist's convention in which the free energy is an upper bound on the negative log evidence. However, other disciplines (including statistics and machine learning) use the negative free energy as an evidence lower bound (or ELBO). These are completely equivalent but can cause some confusion in interdisciplinary research.

5. MAP estimates are the most probable states considering prior beliefs and the data available; contrast this with *maximum likelihood* approaches which do not take prior beliefs into account.

Chapter 5

1. This nomenclature comes from reinforcement learning theories (Daw et al. 2005) but is slightly misleading as both systems depend on models. "Model-free" systems just use a simpler model that predicts a certain kind of behavior in a certain kind of environment.

Chapter 6

1. This does not imply discrete temporal dynamics from a neural perspective. Instead, continuous neural dynamics are seen as representing (continuous) changes in beliefs about (discrete) sequences of events.

2. Having said this, the use of generalized coordinates of motion (box 4.2) in continuous-time models means that they are temporally deep in virtue of their implicit representation of a short trajectory. However, these models do not (necessarily) include variables representing alternative trajectories one could pursue (i.e., the consequences of sequences of actions).

3. Do not confuse temporally deep models with hierarchical models. Unlike temporally deep models, some hierarchical models (e.g., predictive coding models; see section 4.4.1) only consider present observations. However, generative models can be both hierarchical and temporally deep to afford multiscale planning.

Chapter 8

1. Often it is necessary to add damping terms to account for friction and/or viscosity to preclude oscillatory solutions.

Chapter 9

1. Practically, it is often useful to define parameters as log scaling parameters: the parameter acts as a nonnegative scaling factor and cannot be characterized by a normal distribution, which allocates negative numbers a finite probability density. Assuming instead that the log of the scaling parameter is normally distributed ensures positivity when exponentiated to get the scaling parameter itself. The same aim may be achieved by modeling the square root of a parameter as being normally distributed.

2. For example, $\partial_x f(x) \approx \frac{1}{2\Delta x}\big(f(x + \Delta x) - f(x - \Delta x)\big)$.

Chapter 10

1. From a more pragmatic viewpoint, Active Inference only requires the acquisition of forward models, which are (typically) easier to learn compared to inverse models because they are simply a direct (observable) mapping between actions and consequences. Forward models can also be acquired by imitation or external supervision—a technique largely analogous to Active Inference that is widely used to train robotic models (Nishimoto and Tani 2009).

2. In machine learning, the process of optimizing sequences of actions is sometimes called *sequential policy optimization*—as opposed to the more usual *optimization of state-action policies*—namely, "If I am in this state, what do I do?"

3. The notion of deploying cognitive resources efficiently is an inherent part of free energy minimization because minimizing complexity automatically maximizes efficiency, in both an information theoretic and thermodynamic sense. Put simply, the path of least resistance is the path of least free energy.

Appendix A

1. Tensors are a generalization of the concepts of scalars, vectors, and matrices. Heuristically, we can think of these as arrays whose elements are addressable by a certain number of indices. For a vector, we need only a single (row) argument to specify an element. This makes it a first order tensor. For a matrix, we need to specify a column and a row, making it a second order tensor. Scalars need no indices to specify an element, so are 0 order.

2. This uses the identity $\partial_A \ln |A| = A^{-1}$.

3. In the context of variational inference, the integral is typically an expectation.

4. This is sometimes referred to as *the fundamental lemma of variational calculus*.

5. This uses the chain rule, as applied to the derivative of a log: $\partial_x \ln f(x) = f(x)^{-1}\partial_x f(x)$.

Appendix B

1. A normalized exponential function.

2. For concision, we have omitted some terms in the derivatives of the log partition functions. We are licensed to do so by the choice of variational distribution, as any higher order polynomial terms would violate the form of this distribution.

3. The C here is a covariance and should not be confused with a prior preference, despite the same notation in preceding sections.

References

Ackley, D. H., G. E. Hinton, and T. J. Sejnowski (1985). "A learning algorithm for Boltzmann machines." *Cognitive Science* 9(1): 147–169.

Adams, R. A., E. Aponte, L. Marshall, and K. J. Friston (2015). "Active inference and oculomotor pursuit: the dynamic causal modelling of eye movements." *Journal of Neuroscience Methods* 242: 1–14.

Adams, R. A., M. Bauer, D. Pinotsis, and K. J. Friston (2016). "Dynamic causal modelling of eye movements during pursuit: confirming precision-encoding in V1 using MEG." *NeuroImage* 132: 175–189.

Adams, R. A., L. U. Perrinet, and K. Friston (2012). "Smooth pursuit and visual occlusion: Active Inference and oculomotor control in schizophrenia." *PLOS ONE* 7(10): e47502.

Adams, R. A., S. Shipp, and K. J. Friston (2013). "Predictions not commands: Active Inference in the motor system." *Brain Structure and Function* 218(3): 611–643.

Adams, R. A., K. E. Stephan, H. R. Brown, C. D. Frith, and K. J. Friston (2013). "The computational anatomy of psychosis." *Frontiers in Psychiatry* 4: 47.

Aghajanian, G. K., and G. J. Marek (1999). "Serotonin, via 5-HT2A receptors, increases EPSCs in layer V pyramidal cells of prefrontal cortex by an asynchronous mode of glutamate release." *Brain Research* 825(1): 161–171.

Ahmadi, A., and J. Tani (2019). "A novel predictive-coding-inspired variational RNN model for online prediction and recognition." *Neural Computation* 31(11): 2025–2074.

Aitchison, L., and M. Lengyel (2017). "With or without you: predictive coding and Bayesian inference in the brain." *Current Opinion in Neurobiology* 46: 219–227.

Allen, M., A. Levy, T. Parr, and K. J. Friston (2019). "In the body's eye: the computational anatomy of interoceptive inference." *bioRxiv* 603928.

Anderson, B. A., P. A. Laurent, and S. Yantis (2011). "Value-driven attentional capture." *Proceedings of the National Academy of Sciences* 108(25): 10367.

Arnal, L. H., and A.-L. Giraud (2012). "Cortical oscillations and sensory predictions." *Trends in Cognitive Sciences* 16(7): 390–398.

Arnsten, A. F. T., and B.-M. Li (2005). "Neurobiology of executive functions: catecholamine influences on prefrontal cortical functions." *Biological Psychiatry* 57(11): 1377–1384.

Ashby, W. R. (1952). *Design for a Brain*. Oxford: Wiley.

Attias, H. (2003). "Planning by probabilistic inference." *Proceedings of the 9th International Workshop on Artificial Intelligence and Statistics, Key West, Florida, USA.*

Baldassarre, G., and M. Mirolli (2013). *Intrinsically Motivated Learning in Natural and Artificial Systems*. New York: Springer.

Barca, L., and G. Pezzulo (2020). "Keep your interoceptive streams under control: an Active Inference perspective on anorexia nervosa." *Cognitive, Affective and Behavioral Neuroscience* 20(2): 427–440.

Barrett, L. F. (2017). *How Emotions Are Made: The Secret Life of the Brain*. Boston, MA: Houghton Mifflin Harcourt.

Barrett, L. F., K. S. Quigley, and P. Hamilton (2016). "An Active Inference theory of allostasis and interoception in depression." *Philosophical Transactions of the Royal Society B* 371(1708): 20160011.

Barrett, L. F., and W. K. Simmons (2015). "Interoceptive predictions in the brain." *Nature Reviews Neuroscience* 16(7): 419–429.

Barsalou, L. W. (2008). "Grounded cognition." *Annual Review of Psychology* 59: 617–645.

Bastos, A. M., V. Litvak, R. Moran, C. A. Bosman, P. Fries, and K. J. Friston (2015). "A DCM study of spectral asymmetries in feedforward and feedback connections between visual areas V1 and V4 in the monkey." *Neuroimage* 108: 460–475.

Bastos, A. M., W. M. Usrey, R. A. Adams, G. R. Mangun, P. Fries, and K. J. Friston (2012). "Canonical microcircuits for predictive coding." *Neuron* 76(4): 695–711.

Beal, M. J. (2003). "Variational algorithms for approximate Bayesian inference." PhD diss., University of London.

Bellman, R. (1954). "The theory of dynamic programming." *Bulletin of the American Mathematical Society* 60(6): 503–515.

Benrimoh, D., T. Parr, P. Vincent, R. A. Adams, and K. Friston (2018). "Active Inference and auditory hallucinations." *Computational Psychiatry* 2: 183–204.

Berridge, K. C. (2007). "The debate over dopamine's role in reward: the case for incentive salience." *Psychopharmacology* 191(3): 391–431.

Berridge, K. C., and M. L. Kringelbach (2011). "Building a neuroscience of pleasure and well-being." *Psychology of Well-Being* 1(1): 1–3.

Botvinick, M., and M. Toussaint (2012). "Planning as inference." *Trends in Cognitive Sciences* 16(10): 485–488.

Botvinick, M. M. (2008). "Hierarchical models of behavior and prefrontal function." *Trends in Cognitive Sciences* 12(5): 201–208.

Brown, H., R. A. Adams, I. Parees, M. Edwards, and K. Friston (2013). "Active Inference, sensory attenuation and illusions." *Cognitive Processing* 14(4): 411–427.

Brown, H., and K. Friston (2012). "Free-energy and illusions: the cornsweet effect." *Frontiers in Psychology* 3(43).

Brown, L. D. (1981). "A complete class theorem for statistical problems with finite-sample spaces." *Annals of Statistics* 9(6): 1289–1300.

Bruineberg, J., J. Kiverstein, and E. Rietveld (2016). "The anticipating brain is not a scientist: the free-energy principle from an ecological-enactive perspective." *Synthese* 195: 2417–2444.

Bruineberg, J., E. Rietveld, T. Parr, L. van Maanen, and K. J. Friston (2018). "Free-energy minimization in joint agent-environment systems: a niche construction perspective." *Journal of Theoretical Biology* 455: 161–178.

Buzsaki, G. (2019). *The Brain from Inside Out.* New York: Oxford University Press.

Callaway, E. M., and A. K. Wiser (2009). "Contributions of individual layer 2–5 spiny neurons to local circuits in macaque primary visual cortex." *Visual Neuroscience* 13(5): 907–922.

Cannon, W. B. (1929). "Organization for physiological homeostasis." *Physiological Reviews* 9(3): 399–431.

Ciria, A., G. Schillaci, G. Pezzulo, V. V. Hafner, and B. Lara (2021). "Predictive processing in cognitive robotics: a review." *arXiv preprint arXiv:2101.06611.*

Cisek, P. (2019). "Resynthesizing behavior through phylogenetic refinement." *Attention, Perception, and Psychophysics* 81(7): 2265–2287.

Clark, A. (2013). "Whatever next? Predictive brains, situated agents, and the future of cognitive science." *Behavioral and Brain Sciences* 36(03): 181–204.

Clark, A. (2015). *Surfing Uncertainty: Prediction, Action, and the Embodied Mind.* New York: Oxford University Press.

Clark, A., and D. J. Chalmers (1998). "The extended mind." *Analysis* 58: 10–23.

Clark, J. E., S. Watson, and K. J. Friston (2018). "What is mood? A computational perspective." *Psychological Medicine* 48(14): 2277–2284.

Collins, S. H., M. Wisse, and A. Ruina (2016). "A three-dimensional passive-dynamic walking robot with two legs and knees." *International Journal of Robotics Research* 20(7): 607–615.

Conant, R. C., and W. R. Ashby (1970). "Every good regulator of a system must be a model of that system." *International Journal of Systems Science* 1(2): 89–97.

Corcoran, A. W., G. Pezzulo, and J. Hohwy (2020). "From allostatic agents to counterfactual cognisers: Active Inference, biological regulation, and the origins of cognition." *Biology and Philosophy* 35(3): 32.

Corlett, P. R., G. Horga, P. C. Fletcher, B. Alderson-Day, K. Schmack, and A. R. Powers III (2019). "Hallucinations and strong priors." *Trends in Cognitive Sciences* 23(2): 114–127.

Cox, D. R., and H. D. Miller (1965). "The theory of stochastic processes." London: Chapman and Hall/CRC.

Craik, K. (1943). *The Nature of Explanation*. Cambridge: Cambridge University Press.

Cullen, M., B. Davey, K. J. Friston, and R. J. Moran (2018). "Active inference in OpenAI gym: a paradigm for computational investigations into psychiatric illness." *Biological Psychiatry: Cognitive Neuroscience and Neuroimaging* 3(9): 809–818.

Da Costa, L., T. Parr, N. Sajid, S. Veselic, V. Neacsu, and K. Friston (2020). "Active Inference on discrete state-spaces: a synthesis." *arXiv preprint arXiv:2001.07203*.

Daunizeau, J., H. E. M. den Ouden, M. Pessiglione, S. J. Kiebel, K. E. Stephan, and K. J. Friston (2010). "Observing the observer (I): meta-Bayesian models of learning and decision-making." *PLOS ONE* 5(12): e15554.

Dauwels, J. (2007). "On variational message passing on factor graphs." *2007 IEEE International Symposium on Information Theory*, 2546–2550.

Daw, N. D., Y. Niv, and P. Dayan (2005). "Uncertainty-based competition between prefrontal and dorsolateral striatal systems for behavioral control." *Nature Neuroscience* 8(12): 1704–1711.

Dayan, P., G. E. Hinton, R. M. Neal, and R. S. Zemel (1995). "The Helmholtz machine." *Neural Computation* 7: 889–904.

Demirdjian, D., L. Taycher, G. Shakhnarovich, K. Grauman, and T. Darrell (2005). "Avoiding the 'streetlight effect': tracking by exploring likelihood modes." In *Tenth IEEE International Conference on Computer Vision (ICCV'05) Volume 1*, 357–364.

Dennett, D. C. (1978). "Why not the whole iguana?" *Behavioral and Brian Sciences* 1: 103–104.

Dickinson, A., and B. Balleine (1990). "Motivational control of instrumental performance following a shift from thirst to hunger." *Quarterly Journal of Experimental Psychology* 42(4): 413–431.

Disney, A. A., C. Aoki, and M. J. Hawken (2007). "Gain modulation by nicotine in Macaque V1." *Neuron* 56(4): 701–713.

Donnarumma, F., M. Costantini, E. Ambrosini, K. Friston, and G. Pezzulo (2017). "Action perception as hypothesis testing." *Cortex: A Journal Devoted to the Study of the Nervous System and Behavior* 89: 45–60.

Doya, K. (2007). *Bayesian Brain: Probabilistic Approaches to Neural Coding*. Cambridge, MA: MIT Press.

Elliott, M. C., P. M. Tanaka, R. W. Schwark, and R. Andrade (2018). "Serotonin differentially regulates L5 pyramidal cell classes of the medial prefrontal cortex in rats and mice." *eNeuro* 5(1): eneuro.0305–0317.2018.

Feldman, A. G. (2009). "New insights into action-perception coupling." *Experimental Brain Research* 194(1): 39–58.

Feldman, A. G., and M. F. Levin (2009). "The equilibrium-point hypothesis—past, present and future." In *Progress in Motor Control: A Multidisciplinary Perspective*, edited by D. Sternad, 699–726. Boston, MA: Springer US.

Feldman, H., and K. Friston (2010). "Attention, uncertainty, and free-energy." *Frontiers in Human Neuroscience* 4(215).

Felleman, D. J., and D. C. Van Essen (1991). "Distributed hierarchical processing in the primate cerebral cortex." *Cerebral Cortex* 1(1): 1–47.

Fiser, J., P. Berkes, G. Orbán, and M. Lengyel (2010). "Statistically optimal perception and learning: from behavior to neural representations." *Trends in Cognitive Sciences* 14(3): 119–130.

FitzGerald, T. H. B., R. J. Dolan, and K. Friston (2015). "Dopamine, reward learning, and active inference." *Frontiers in Computational Neuroscience* 9: 1–16.

FitzGerald, T. H. B., P. Schwartenbeck, M. Moutoussis, R. J. Dolan, and K. Friston (2015). "Active inference, evidence accumulation, and the urn task." *Neural Computation* 27(2): 306–328.

Foster, D. (2019). *Generative Deep Learning: Teaching Machines to Paint, Write, Compose, and Play*. Boston: O'Reilly Media.

Fountas, Z., N. Sajid, P. A. M. Mediano, and K. Friston (2020). "Deep active inference agents using Monte-Carlo methods." *arXiv:2006.04176 [cs, q-bio, stat]*.

Fradkin, I., R. A. Adams, T. Parr, J. P. Roiser, and J. D. Huppert (2020). "Searching for an anchor in an unpredictable world: a computational model of obsessive compulsive disorder." *Psychological Review* 127(5): 672–699.

Frank, M. J. (2005). "Dynamic dopamine modulation in the basal ganglia: a neurocomputational account of cognitive deficits in medicated and nonmedicated Parkinsonism." *Journal of Cognitive Neuroscience* 17(1): 51–72.

Freeze, B. S., A. V. Kravitz, N. Hammack, J. D. Berke, and A. C. Kreitzer (2013). "Control of basal ganglia output by direct and indirect pathway projection neurons." *Journal of Neuroscience* 33(47): 18531–18539.

Freund, T. F., J. F. Powell, and A. D. Smith (1984). "Tyrosine hydroxylase-immunoreactive boutons in synaptic contact with identified striatonigral neurons, with particular reference to dendritic spines." *Neuroscience* 13(4): 1189–1215.

Friston, K. (2005). "A theory of cortical responses." *Philosophical Transactions of the Royal Society of London B: Biological Sciences* 360(1456): 815–836.

Friston, K. (2008). "Hierarchical models in the brain." *PLOS Computational Biology* 4(11): e1000211.

Friston, K. (2009). "The free-energy principle: a rough guide to the brain?" *Trends in Cognitive Sciences* 13(7): 293–301.

Friston, K. (2011). "What is optimal about motor control?" *Neuron* 72(3): 488–498.

Friston, K. (2013). "Life as we know it." *Journal of the Royal Society Interface* 10(86): 20130475.

Friston, K. (2017). "Precision psychiatry." *Biological Psychiatry: Cognitive Neuroscience and Neuroimaging* 2(8): 640–643.

Friston, K. (2019a). "A free energy principle for a particular physics." *arXiv preprint arXiv:1906.10184.*

Friston, K. (2019b). "Waves of prediction." *PLOS Biology* 17(10): e3000426.

Friston, K., R. Adams, L. Perrinet, and M. Breakspear (2012). "Perceptions as hypotheses: saccades as experiments." *Frontiers in Psychology* 3(151).

Friston, K., and G. Buzsaki (2016). "The functional anatomy of time: what and when in the brain." *Trends in Cognitive Sciences* 20(7): 500–511.

Friston, K., L. Da Costa, D. Hafner, C. Hesp, and T. Parr (2020). "Sophisticated inference." *arXiv preprint arXiv:2006.04120.*

Friston, K., J. Daunizeau, and S. J. Kiebel (2009). "Reinforcement learning or Active Inference?" *PLOS ONE* 4(7): e6421.

Friston, K., J. Daunizeau, J. Kilner, and S. J. Kiebel (2010). "Action and behavior: a free-energy formulation." *Biological Cybernetics* 102(3): 227–260.

Friston, K., T. FitzGerald, F. Rigoli, P. Schwartenbeck, J. O'Doherty, and G. Pezzulo (2016). "Active Inference and learning." *Neuroscience and Biobehavioral Reviews* 68: 862–879.

Friston, K., T. FitzGerald, F. Rigoli, P. Schwartenbeck, and G. Pezzulo (2017). "Active Inference: a process theory." *Neural Computation* 29(1): 1–49.

Friston, K., and C. D. Frith (2015a). "Active inference, communication and hermeneutics( )." *Cortex: A Journal Devoted to the Study of the Nervous System and Behavior* 68: 129–143.

Friston, K., and C. Frith (2015b). "A duet for one." *Consciousness and Cognition* 36: 390–405.

Friston, K., and I. Herreros (2016). "Active Inference and learning in the cerebellum." *Neural Computation* 28(9): 1812–1839.

Friston, K., and S. Kiebel (2009). "Predictive coding under the free-energy principle." *Philosophical Transactions of the Royal Society B: Biological Sciences* 364(1521): 1211.

Friston, K., M. Levin, B. Sengupta, and G. Pezzulo (2015). "Knowing one's place: a free-energy approach to pattern regulation." *Journal of the Royal Society Interface* 12(105): 20141383.

Friston, K., M. Lin, C. D. Frith, G. Pezzulo, J. A. Hobson, and S. Ondobaka (2017). "Active Inference, curiosity and insight." *Neural Computation* 29(10): 2633–2683.

Friston, K., V. Litvak, A. Oswal, A. Razi, K. E. Stephan, B. C. M. van Wijk, G. Ziegler, and P. Zeidman (2016). "Bayesian model reduction and empirical Bayes for group (DCM) studies." *NeuroImage* 128(Supplement C): 413–431.

Friston, K., J. Mattout, and J. Kilner (2011). "Action understanding and active inference." *Biological Cybernetics* 104(1): 137–160.

Friston, K., J. Mattout, N. Trujillo-Barreto, J. Ashburner, and W. Penny (2007). "Variational free energy and the Laplace approximation." *NeuroImage* 34(1): 220–234.

Friston, K., T. Parr, and B. de Vries (2017). "The graphical brain: belief propagation and Active Inference." *Network Neuroscience* 1(4): 381–414.

Friston, K., T. Parr, Y. Yufik, N. Sajid, C. J. Price, and E. Holmes (2020). "Generative models, linguistic communication and active inference." *Neuroscience and Biobehavioral Reviews* 118: 42–64.

Friston, K., T. Parr, and P. Zeidman (2018). "Bayesian model reduction." *arXiv preprint arXiv:1805.07092*.

Friston, K., F. Rigoli, D. Ognibene, C. Mathys, T. Fitzgerald, and G. Pezzulo (2015). "Active Inference and epistemic value." *Cognitive Neuroscience* 6(4): 187–214.

Friston, K., R. Rosch, T. Parr, C. Price, and H. Bowman (2017). "Deep temporal models and active inference." *Neuroscience and Biobehavioral Reviews* 77: 388–402.

Friston, K., S. Samothrakis and R. Montague (2012). "Active Inference and agency: optimal control without cost functions." *Biological Cybernetics* 106(8–9): 523–541.

Friston, K., P. Schwartenbeck, T. FitzGerald, M. Moutoussis, T. Behrens, and R. J. Dolan (2014). "The anatomy of choice: dopamine and decision-making." *Philosophical Transactions of the Royal Society B: Biological Sciences* 369(1655): 20130481.

Friston, K., K. Stephan, B. Li, and J. Daunizeau (2010). "Generalised filtering." *Mathematical Problems in Engineering*. doi:10.1155/2010/621670.

Friston, K., K. E. Stephan, R. Montague, and R. J. Dolan (2014). "Computational psychiatry: the brain as a phantastic organ." *Lancet Psychiatry* 1(2): 148–158.

Frith, C. D., S. Blakemore, and D. M. Wolpert (2000). "Explaining the symptoms of schizophrenia: abnormalities in the awareness of action." *Brain Research Reviews* 31(2–3): 357–363.

Funahashi, S., C. J. Bruce, and P. S. Goldman-Rakic (1989). "Mnemonic coding of visual space in the monkey's dorsolateral prefrontal cortex." *Journal of Neurophysiology* 61(2): 331.

Fuster, J. n. M. (2004). "Upper processing stages of the perception-action cycle." *Trends in Cognitive Sciences* 8(4): 143–145.

Galea, J. M., S. Bestmann, M. Beigi, M. Jahanshahi, and J. C. Rothwell (2012). "Action reprogramming in Parkinson's disease: response to prediction error is modulated by levels of dopamine." *Journal of Neuroscience* 32(2): 542.

George, D., W. Lehrach, K. Kansky, M. Lázaro-Gredilla, C. Laan, B. Marthi, X. Lou, Z. Meng, Y. Liu, H. Wang, A. Lavin, and D. S. Phoenix (2017). "A generative vision model that trains with high data efficiency and breaks text-based CAPTCHAs." *Science* 358(6368): eaag2612.

Gershman, S. J., E. J. Horvitz, and J. B. Tenenbaum (2015). "Computational rationality: a converging paradigm for intelligence in brains, minds, and machines." *Science* 349(6245): 273.

Gertler, T. S., C. S. Chan, and D. J. Surmeier (2008). "Dichotomous anatomical properties of adult striatal medium spiny neurons." *Journal of Neuroscience* 28(43): 10814.

Gil, Z., B. W. Connors, and Y. Amitai (1997). "Differential regulation of neocortical synapses by neuromodulators and activity." *Neuron* 19(3): 679–686.

Goodfellow, I. J., J. Pouget-Abadie, M. Mirza, B. Xu, D. Warde-Farley, S. Ozair, A. Courville, and Y. Bengio (2014). "Generative adversarial networks." *arXiv:1406.2661 [cs, stat]*.

Gottlieb, J., P.-Y. Oudeyer, M. Lopes, and A. Baranes (2013). "Information-seeking, curiosity, and attention: computational and neural mechanisms." *Trends in Cognitive Sciences* 17(11): 585–593.

Gottwald, S., and D. A. Braun (2020). "The two kinds of free energy and the Bayesian revolution." *arXiv:2004.11763 [cs, q-bio]*.

Gregory, R. L. (1980). "Perceptions as hypotheses." *Philosophical Transactions of the Royal Society of London B: Biological Sciences* 290(1038): 181–197.

Ha, D., and D. Eck (2017). "A neural representation of sketch drawings." *arXiv pre-print arXiv:1704.03477.*

Ha, D., and J. Schmidhuber (2018). "World models." *arXiv:1803.10122 [cs, stat].*

Haeusler, S., and W. Maass (2007). "A statistical analysis of information-processing properties of lamina-specific cortical microcircuit models." *Cerebral Cortex* 17(1): 149–162.

Harlow, H. F. (1949). "The formation of learning sets." *Psychological Review* 56(1): 51–65.

Helmholtz, H. v. (1866). "Concerning the perceptions in general." *Treatise on Physiological Optics.* Translated by J. P. C. Southall. New York, Dover.

Helmholtz, H. v. (1867). *Handbuch der physiologischen Optik.* Leipzig: L. Voss.

Herbart, J. (1825). *Psychologie als Wissenschaft: Neu gegründet auf Erfahrung, Metaphysik und Mathematik.* Zweiter, analytischer Teil. Koenigsberg, Germany: August Wilhem Unzer.

Hezemans, F. H., N. Wolpe, and J. B. Rowe (2020). "Apathy is associated with reduced precision of prior beliefs about action outcomes." *Journal of Experimental Psychology: General* 149(9): 1767–1777.

Hills, T. T., P. M. Todd, D. Lazer, A. D. Redish, and I. D. Couzin (2015). "Exploration versus exploitation in space, mind, and society." *Trends in Cognitive Sciences* 19(1): 46–54.

Hillyard, S. A., E. K. Vogel, and S. J. Luck (1998). "Sensory gain control (amplification) as a mechanism of selective attention: electrophysiological and neuroimaging evidence." *Philosophical Transactions of the Royal Society B: Biological Sciences* 353(1373): 1257–1270.

Hinton, G. E. (2007a). "Learning multiple layers of representation." *Trends in Cognitive Sciences* 11(10): 428–434.

Hinton, G. E. (2007b). "To recognize shapes, first learn to generate images." *Progress in Brain Research* 165: 535–547.

Hoffmann, J. (1993). *Vorhersage und Erkenntnis: Die Funktion von Antizipationen in der menschlichen Verhaltenssteuerung und Wahrnehmung* [Anticipation and cognition: The function of anticipations in human behavioral control and perception]. Goettingen, Germany: Hogrefe.

Hoffmann, J. (2003). "Anticipatory behavioral control." In *Anticipatory Behavior in Adaptive Learning Systems: Foundations, Theories, and Systems,* edited by M. V. Butz, O. Sigaud, and P. Gerard, 44–65. Berlin: Springer-Verlag.

Hohwy, J. (2013). *The Predictive Mind.* New York: Oxford University Press.

Hohwy, J. (2016). "The self-evidencing brain." *Noûs* 50(2): 259–285.

Hommel, B., J. Musseler, G. Aschersleben, and W. Prinz (2001). "The theory of event coding (TEC): a framework for perception and action planning." *Behavioral and Brain Science* 24(5): 849–878.

Huerta, R., and M. Rabinovich (2004). "Reproducible sequence generation in random neural ensembles." *Physical Review Letters* 93(23): 238104.

Hurley, S. (2008). "The shared circuits model (SCM): how control, mirroring, and simulation can enable imitation, deliberation, and mindreading." *Behavioral and Brain Sciences* 31: 1–22.

Huygens, C. (1673). *Horologium Oscillatorium: Sive, De Motu Pendulorum Ad Horologia Aptato Demostrationes Geometricae.* Culture et Civilisation.

Iodice, P., G. Porciello, I. Bufalari, L. Barca, and G. Pezzulo (2019). "An interoceptive illusion of effort induced by false heart-rate feedback." *Proceedings of the National Academy of Sciences* 116(28): 13897–13902.

Isomura, T., and K. Friston (2018). "In vitro neural networks minimise variational free energy." *Scientific Reports* 8(1): 16926.

Isomura, T., T. Parr, and K. Friston (2019). "Bayesian filtering with multiple internal models: toward a theory of social intelligence." *Neural Computation* 31(12): 2390–2431.

James, W. (1890). *The Principles of Psychology.* New York: Dover Publications.

Jaynes, E. T. (1957). "Information theory and statistical mechanics." *Physical Review* 106(4): 620.

Jeannerod, M. (2001). "Neural simulation of action: a unifying mechanism for motor cognition." *NeuroImage* 14: S103–S109.

Joffily, M., and G. Coricelli (2013). "Emotional valence and the free-energy principle." *PLOS Computational Biology* 9(6): e1003094.

Kahneman, D. (2017). Thinking, fast and slow. United Kingdom: Penguin Books.

Kakade, S., and P. Dayan (2002). "Dopamine: generalization and bonuses." *Neural Networks* 15(4): 549–559.

Kanai, R., Y. Komura, S. Shipp, and K. Friston (2015). "Cerebral hierarchies: predictive processing, precision and the pulvinar." *Philosophical Transactions of the Royal Society B: Biological Sciences* 370(1668): 20140169.

Kaplan, R., and K. J. Friston (2018). "Planning and navigation as Active Inference." *Biological Cybernetics* 112: 323–343.

Kappen, H. J., V. Gómez, and M. Opper (2012). "Optimal control as a graphical model inference problem." *Machine Learning* 87(2): 159–182.

Karson, C. N. (1983). "Spontaneous eye-blink rates and dopaminergic systems." *Brain* 106(3): 643–653.

Kemp, C., and J. B. Tenenbaum (2008). "The discovery of structural form." *Proceedings of the National Academy of Sciences* 105(31): 10687–10692.

Kiebel, S. J., J. Daunizeau, and K. J. Friston (2008). "A hierarchy of time-scales and the brain." *PLOS Computational Biology* 4(11): e1000209.

Kingma, D. P., and M. Welling (2014). "Auto-encoding variational Bayes." *arXiv:1312.6114 [cs, stat].*

Kirchhoff, M., T. Parr, E. Palacios, K. Friston, and J. Kiverstein (2018). "The Markov blankets of life: autonomy, active inference and the free energy principle." *Journal of the Royal Society*, Interface 15, 20170792, doi:10.1098/rsif.2017.0792.

Knill, D. C., and A. Pouget (2004). "The Bayesian brain: The role of uncertainty in neural coding and computation." *Trends in Neurosciences* 27(12): 712–719.

Kording, K. P., and D. M. Wolpert (2006). "Bayesian decision theory in sensorimotor control." *Trends in Cognitive Sciences* 10: 319–326.

Koss, M. C. (1986). "Pupillary dilation as an index of central nervous system α2-adrenoceptor activation." *Journal of Pharmacological Methods* 15(1): 1–19.

Krakauer, J. W., A. A. Ghazanfar, A. Gomez-Marin, M. A. MacIver, and D. Poeppel (2017). "Neuroscience needs behavior: correcting a reductionist bias." *Neuron* 93(3): 480–490.

Krishnamurthy, K., M. R. Nassar, S. Sarode, and J. I. Gold (2017). "Arousal-related adjustments of perceptual biases optimize perception in dynamic environments." *Nature Human Behaviour* 1: 0107.

Kunde, W., I. Koch, and J. Hoffmann (2004). "Anticipated action effects affect the selection, initiation and execution of actions." *Quarterly Journal of Experimental Psychology. Section A: Human Experimental Psychology* 57(1): 87–106.

Lake, B. M., T. D. Ullman, J. B. Tenenbaum, and S. J. Gershman (2017). "Building machines that learn and think like people." *Behavioral and Brain Sciences* 40: 1–72.

Lambe, E. K., P. S. Goldman-Rakic, and G. K. Aghajanian (2000). "Serotonin induces EPSCs preferentially in layer V pyramidal neurons of the frontal cortex in the rat." *Cerebral Cortex* 10(10): 974–980.

Lavín, C., R. San Martín, and E. Rosales Jubal (2013). "Pupil dilation signals uncertainty and surprise in a learning gambling task." *Frontiers in Behavioral Neuroscience* 7: 218.

Lavine, N., M. Reuben, and P. Clarke (1997). "A population of nicotinic receptors is associated with thalamocortical afferents in the adult rat: laminal and areal analysis." *Journal of Comparative Neurology* 380(2): 175–190.

Lee, M. D., and E.-J. Wagenmakers (2014). *Bayesian Cognitive Modeling: A Practical Course*. Cambridge: Cambridge University Press.

Lee, S. W., S. Shimojo, and J. P. O'Doherty (2014). "Neural computations underlying arbitration between model-based and model-free learning." *Neuron* 81(3): 687–699.

Levine, S. (2018). "Reinforcement learning and control as probabilistic inference: tutorial and review." *arXiv:1805.00909 [cs, stat]*.

Liao, H.-I., M. Yoneya, S. Kidani, M. Kashino, and S. Furukawa (2016). "Human pupillary dilation response to deviant auditory stimuli: effects of stimulus properties and voluntary attention." *Frontiers in Neuroscience* 10: 43.

Limanowski, J., and K. Friston (2019). "Attentional modulation of vision versus proprioception during action." *Cerebral Cortex* 30(3): 1637–1648.

Lindley, D. V. (1956). "On a measure of the information provided by an experiment." *Annals of Mathematical Statistics* 27(4): 986–1005.

Linson, A., T. Parr, and K. J. Friston (2020). "Active Inference, stressors, and psychological trauma: a neuroethological model of (mal)adaptive explore-exploit dynamics in ecological context." *Behavioural Brain Research* 380: 1–13.

Loeliger, H. A. (2004). "An introduction to factor graphs." *IEEE Signal Processing Magazine* 21(1): 28–41.

Loeliger, H. A., J. Dauwels, J. Hu, S. Korl, L. Ping, and F. R. Kschischang (2007). "The factor graph approach to model-based signal processing." *Proceedings of the IEEE* 95(6): 1295–1322.

MacKay, D. M. (1956). *The Epistemological Problem for Automata*. Princeton, NJ: Princeton University Press.

Maisto, D., L. Barca, O. V. d. Bergh, and G. Pezzulo (2021). "Perception and misperception of bodily symptoms from an Active Inference perspective: modelling the case of panic disorder." *Psychological Review*.

Maisto, D., K. Friston, and G. Pezzulo (2019). "Caching mechanisms for habit formation in Active Inference." *Neurocomputing* 359: 298–314.

Marek, R., C. Strobel, T. W. Bredy, and P. Sah (2013). "The amygdala and medial prefrontal cortex: partners in the fear circuit." *Journal of Physiology* 591(10): 2381–2391.

Marshall, L., C. Mathys, D. Ruge, A. O. de Berker, P. Dayan, K. E. Stephan, and S. Bestmann (2016). "Pharmacological fingerprints of contextual uncertainty." *PLOS Biology* 14(11): e1002575.

Maturana, H. R., and F. J. Varela (1980). *Autopoiesis and Cognition: The Realization of Living*. Dordrecht, Holland: D. Reidel.

Mesulam, M. M. (1998). "From sensation to cognition." *Brain: Journal of Neurology* 121(pt. 6): 1013–1052.

Miller, E. K., and J. D. Cohen (2001). "An integrative theory of prefrontal cortex function." *Annual Review of Neuroscience* 24: 167–202.

Miller, G. A., E. Galanter, and K. H. Pribram (1960). *Plans and the Structure of Behavior*. New York: Holt, Rinehart and Winston.

Miller, K. D. (2003). "Understanding layer 4 of the cortical circuit: a model based on cat V1." *Cerebral Cortex* 13(1): 73–82.

Millidge, B. (2019). "Deep Active Inference as variational policy gradients." *arXiv:1907.03876 [cs]*.

Mirza, M. B., R. A. Adams, C. Mathys, and K. J. Friston (2018). "Human visual exploration reduces uncertainty about the sensed world." *PLOS ONE* 13(1): e0190429.

Mirza, M. B., R. A. Adams, C. D. Mathys, and K. J. Friston (2016). "Scene construction, visual foraging, and Active Inference." *Frontiers in Computational Neuroscience* 10(56).

Mirza, M. B., R. A. Adams, T. Parr, and K. Friston (2019). "Impulsivity and Active Inference." *Journal of Cognitive Neuroscience* 31(2): 202–220.

Montague, P. R., R. J. Dolan, K. J. Friston, and P. Dayan (2012). "Computational psychiatry." *Trends in Cognitive Sciences* 16(1): 72–80.

Moran, R. J., P. Campo, M. Symmonds, K. E. Stephan, R. J. Dolan, and K. J. Friston (2013). "Free energy, precision and learning: the role of cholinergic neuromodulation." *Journal of Neuroscience* 33(19): 8227–8236.

Moss, J., and J. P. Bolam (2008). "A dopaminergic axon lattice in the striatum and its relationship with cortical and thalamic terminals." *Journal of Neuroscience* 28(44): 11221.

Moutoussis, M., N. J. Trujillo-Barreto, W. El-Deredy, R. J. Dolan, and K. J. Friston (2014). "A formal model of interpersonal inference." *Frontiers in Human Neuroscience* 8: 160.

Mukherjee, P., A. Sabharwal, R. Kotov, A. Szekely, R. Parsey, D. M. Barch, and A. Mohanty (2016). "Disconnection between amygdala and medial prefrontal cortex in psychotic disorders." *Schizophrenia Bulletin* 42(4): 1056–1067.

Murphy, K. P. (2012). *Machine Learning: A Probabilistic Perspective*. Cambridge, MA: MIT Press.

Nambu, A. (2004). "A new dynamic model of the cortico-basal ganglia loop." *Progress in Brain Research* 143: 461–466.

Nassar, M. R., K. M. Rumsey, R. C. Wilson, K. Parikh, B. Heasly, and J. I. Gold (2012). "Rational regulation of learning dynamics by pupil-linked arousal systems." *Nature Neuroscience* 15(7): 1040–1046.

Nave, K., G. Deane, M. Miller, and A. Clark (2020). "Wilding the predictive brain." *Cognitive Science* 11(6): e1542.

Neisser, U. (2014). *Cognitive Psychology: Classic Edition.* London: Taylor & Francis.

Nishimoto, R., and J. Tani (2009). "Development of hierarchical structures for actions and motor imagery: a constructivist view from synthetic neuro-robotics study." *Psychological Research PRPF* 73(4): 545–558.

Olsen, S. R., D. S. Bortone, H. Adesnik, and M. Scanziani (2012). "Gain control by layer six in cortical circuits of vision." *Nature* 483: 47.

Ortega, P. A., and D. A. Braun (2013). "Thermodynamics as a theory of decision-making with information-processing costs." *Proceedings of the Royal Society A: Mathematical, Physical and Engineering Science* 469(2153).

Oudeyer, P. Y., F. Kaplan, and V. Hafner (2007). "Intrinsic motivation systems for autonomous mental development." *IEEE Transactions on Evolutionary Computation* 11(2): 265–286.

Palacios, E. R., T. Isomura, T. Parr, and K. Friston (2019). "The emergence of synchrony in networks of mutually inferring neurons." *Scientific Reports* 9(1): 6412.

Palacios, E. R., A. Razi, T. Parr, M. Kirchhoff, and K. Friston (2020). "On Markov blankets and hierarchical self-organisation." *Journal of Theoretical Biology* 486: 110089.

Pareés, I., H. Brown, A. Nuruki, R. A. Adams, M. Davare, K. P. Bhatia, K. Friston, and M. J. Edwards (2014). "Loss of sensory attenuation in patients with functional (psychogenic) movement disorders." *Brain* 137(11): 2916–2921.

Parr, T. (2020). "Inferring what to do (and what not to)." *Entropy* 22(5): 536.

Parr, T., D. A. Benrimoh, P. Vincent, and K. J. Friston (2018). "Precision and false perceptual inference." *Frontiers in Integrative Neuroscience* 12: 39–39.

Parr, T., L. D. Costa, and K. Friston (2020). "Markov blankets, information geometry and stochastic thermodynamics." *Philosophical Transactions of the Royal Society A: Mathematical, Physical and Engineering Sciences* 378(2164): 20190159.

Parr, T., and K. J. Friston (2017a). "The computational anatomy of visual neglect." *Cerebral Cortex* 28: 1–14.

Parr, T., and K. J. Friston (2017b). "Uncertainty, epistemics and active inference." *Journal of the Royal Society Interface* 14(136).

Parr, T., and K. J. Friston (2017c). "Working memory, attention, and salience in Active Inference." *Scientific Reports* 7(1): 14678.

Parr, T., and K. J. Friston (2018a). "Active Inference and the anatomy of oculomotion." *Neuropsychologia* 111: 334–343.

Parr, T., and K. J. Friston (2018b). "The anatomy of inference: generative models and brain structure." *Frontiers in Computational Neuroscience* 12(90).

Parr, T., and K. J. Friston (2018c). "The discrete and continuous brain: From decisions to movement—and back again." *Neural Computation* 30(9): 2319–2347.

Parr, T., and K. J. Friston (2018d). "Generalised free energy and Active Inference: can the future cause the past?" *bioRxiv*.

Parr, T., and K. J. Friston (2019a). "Attention or salience?" *Current Opinion in Psychology* 29: 1–5.

Parr, T., and K. J. Friston (2019b). "The computational pharmacology of oculomotion." *Psychopharmacology* 236(8): 2473–2484.

Parr, T., D. Markovic, S. J. Kiebel, and K. J. Friston (2019). "Neuronal message passing using mean-field, Bethe, and marginal approximations." *Scientific Reports* 9(1): 1889.

Parr, T., M. B. Mirza, H. Cagnan, and K. J. Friston (2019). "Dynamic causal modelling of active vision." *Journal of Neuroscience* 39(32): 6265–6275.

Parr, T., R. V. Rikhye, M. M. Halassa, and K. J. Friston (2019). "Prefrontal computation as Active Inference." *Cerebral Cortex* 30(2): 682–695.

Pearl, J. (1988). *Probabilistic Reasoning in Intelligent Systems: Networks of Plausible Inference*. San Francisco, CA: Morgan Kaufmann.

Pearl, J. and D. Mackenzie (2018). *The Book of Why: The New Science of Cause and Effect*. New York: Basic Books.

Perrinet, L. U., R. A. Adams, and K. J. Friston (2014). "Active Inference, eye movements and oculomotor delays." *Biological Cybernetics* 108(6): 777–801.

Peters, A., B. S. McEwen, and K. Friston (2017). "Uncertainty and stress: why it causes diseases and how it is mastered by the brain." *Progress in Neurobiology*. 156: 164–188

Petersen, K. B., and M. S. Pedersen (2012). *The Matrix Cookbook*. https://www.math .uwaterloo.ca/~hwolkowi/matrixcookbook.pdf.

Pezzulo, G. (2012). "An Active Inference view of cognitive control." *Frontiers in Theoretical and Philosophical Psychology* 478: 1–2

Pezzulo, G. (2013). "Why do you fear the bogeyman? An embodied predictive coding model of perceptual inference." *Cognitive, Affective, and Behavioral Neuroscience* 14(3): 902–911.

Pezzulo, G., G. Baldassarre, M. V. Butz, C. Castelfranchi, and J. Hoffmann (2007). "Fron action to goals and vice-versa: Theoretical analysis and models of the ideomotor principle and TOTE." In *Anticipatory Behavior in Adaptive Learning Systems*, edited by M. V. Butz, O. Sigaud, G. Pezzulo, and G. Baldassarre. ABiALS 2006. *Lecture Notes in Computer Science*, vol 4520. Berlin: Springer. https://doi.org/10.1007/978-3-540-74262-3_5.

Pezzulo, G., L. W. Barsalou, A. Cangelosi, M. H. Fischer, K. McRae, and M. J. Spivey (2013). "Computational grounded cognition: a new alliance between grounded cognition and computational modeling." *Frontiers in Psychology* 3: 612.

Pezzulo, G., E. Cartoni, F. Rigoli, L. Pio-Lopez, and K. Friston (2016). "Active Inference, epistemic value, and vicarious trial and error." *Learning and Memory* 23(7): 322–338.

Pezzulo, G., and P. Cisek (2016). "Navigating the affordance landscape: feedback control as a process model of behavior and cognition." *Trends in Cognitive Sciences* 20(6): 414–424.

Pezzulo, G., F. Donnarumma, P. Iodice, D. Maisto, and I. Stoianov (2017). "Model-based approaches to active perception and control." *Entropy* 19(6): 266.

Pezzulo, G., C. Kemere, and M. A. A. van der Meer (2017). "Internally generated hippocampal sequences as a vantage point to probe future-oriented cognition." *Annals of the New York Academy of Sciences* 1396(1): 144–165.

Pezzulo, G., and M. Levin (2015). "Re-membering the body: applications of computational neuroscience to the top-down control of regeneration of limbs and other complex organs." *Integrative Biology* 7(12): 1487–1517.

Pezzulo, G., B. Lw, A. Cangelosi, M. H. Fischer, K. McRae, and M. Spivey (2011). "The mechanics of embodiment: A dialogue on embodiment and computational modeling." *Frontiers in Cognition* 2(5): 1–21.

Pezzulo, G., D. Maisto, L. Barca, and O. V. d. Bergh (2019). "Symptom perception from a predictive processing perspective." *Clinical Psychology in Europe* 1(4): 1–14.

Pezzulo, G., and F. Rigoli (2011). "The value of foresight: how prospection affects decision-making." *Frontiers in Neuroscience* 5(79).

Pezzulo, G., F. Rigoli, and K. J. Friston (2015). "Active Inference, homeostatic regulation and adaptive behavioural control." *Progress in Neurobiology* 136: 17–35.

Pezzulo, G., F. Rigoli, and K. J. Friston (2018). "Hierarchical Active Inference: a theory of motivated control." *Trends in Cognitive Sciences* 22(4): 294–306.

Pezzulo, G., M. Zorzi, and M. Corbetta (2020). "The secret life of predictive brains: what's spontaneous activity for?" *psyarxiv*.

Pfeifer, R., and J. C. Bongard (2006). *How the Body Shapes the Way We Think*. Cambridge, MA: MIT Press.

Pio-Lopez, L., A. Nizard, K. Friston, and G. Pezzulo (2016). "Active Inference and robot control: a case study." *Journal of the Royal Society Interface* 13(122).

Posner, M. I., R. D. Rafal, L. S. Choate, and J. Vaughan (1985). "Inhibition of return: neural basis and function." *Cognitive Neuropsychology* 2(3): 211–228.

Pouget, A., J. M. Beck, W. J. Ma, and P. E. Latham (2013). "Probabilistic brains: knowns and unknowns." *Nature Neuroscience* 16(9): 1170–1178.

Powers, W. T. (1973). *Behavior: The Control of Perception*. Hawthorne, NY: Aldine.

Prosser, A., K. J. Friston, N. Bakker, and T. Parr (2018). "A Bayesian account of psychopathy: a model of lacks remorse and self-aggrandizing." *Computational Psychiatry* 1–49.

Ramstead, M. J. D., M. D. Kirchhoff, and K. J. Friston (2019). "A tale of two densities: Active Inference is enactive inference." *Adaptive Behavior* 28(4): 225–239.

Rao, R. P., and D. H. Ballard (1999). "Predictive coding in the visual cortex: a functional interpretation of some extra-classical receptive-field effects." *Nature Neuroscience* 2(1): 79–87.

Rawlik, K., M. Toussaint, and S. Vijayakumar (2013). "On stochastic optimal control and reinforcement learning by approximate inference." In *Robotics: Science and Systems VIII*, edited by N. Roy, P. Newman, and S. Srinivasa. Cambridge, MA: MIT Press.

Risken, H. (1996). "Fokker-Planck equation." *The Fokker-Planck Equation: Methods of Solution and Applications*, 63–95. Berlin: Springer.

Rizzolatti, G., L. Riggio, I. Dascola, and C. Umiltá (1987). "Reorienting attention across the horizontal and vertical meridians: evidence in favor of a premotor theory of attention." *Neuropsychologia* 25(1, pt. 1): 31–40.

Rosenblueth, A., N. Wiener, and J. Bigelow (1943). "Behavior, purpose and teleology." *Philosophy of Science* 10(1): 18–24.

Sahin, M., W. D. Bowen, and J. P. Donoghue (1992). "Location of nicotinic and muscarinic cholinergic and μ-opiate receptors in rat cerebral neocortex: evidence from thalamic and cortical lesions." *Brain Research* 579(1): 135–147.

Sales, A. C., K. J. Friston, M. W. Jones, A. E. Pickering, and R. J. Moran (2019). "Locus coeruleus tracking of prediction errors optimises cognitive flexibility: an Active Inference model." *PLOS Computational Biology* 15(1): e1006267.

Sancaktar, C., M. van Gerven, and P. Lanillos (2020). "End-to-end pixel-based deep Active Inference for body perception and action." *arXiv:2001.05847 [cs, q-bio]*.

Schmidhuber, J. (1991). "Adaptive confidence and adaptive curiosity." Institut fur Informatik, Technische Universitat Munchen.

Schultz, W., P. Dayan, and P. R. Montague (1997). "A neural substrate of prediction and reward." *Science* 275(5306): 1593.

Schwartenbeck, P., T. H. B. FitzGerald, C. Mathys, R. Dolan, and K. Friston (2015). "The dopaminergic midbrain encodes the expected certainty about desired outcomes." *Cerebral Cortex* 25(10): 3434–3445.

Schwartenbeck, P., T. H. B. FitzGerald, C. Mathys, R. Dolan, F. Wurst, M. Kronbichler, and K. Friston (2015). "Optimal inference with suboptimal models: Addiction and active Bayesian inference." *Medical Hypotheses* 84(2): 109–117.

Schwartenbeck, P., and K. Friston (2016). "Computational phenotyping in psychiatry: a worked example." *eNeuro* 3(4): eneuro.0049–0016.2016.

Schwartenbeck, P., J. Passecker, T. U. Hauser, T. H. FitzGerald, M. Kronbichler, and K. J. Friston (2019). "Computational mechanisms of curiosity and goal-directed exploration." *eLife* 8: e41703.

Schwöbel, S., S. Kiebel, and D. Marković (2018). "Active Inference, belief propagation, and the Bethe approximation." *Neural Computation* 30(9): 1–38.

Seth, A. K. (2013). "Interoceptive inference, emotion, and the embodied self." *Trends in Cognitive Sciences* 17(11): 565–573.

Seth, A. K., and K. J. Friston (2016). "Active interoceptive inference and the emotional brain." *Philosophical Transactions of the Royal Society B* 371(1708): 20160007.

Seth, A. K., K. Suzuki, and H. D. Critchley (2012). "An interoceptive predictive coding model of conscious presence." *Frontiers in Psychology* 2: 1–16.

Shadmehr, R., M. A. Smith, and J. W. Krakauer (2010). "Error correction, sensory prediction, and adaptation in motor control." *Annual Review of Neuroscience* 33: 89–108.

Sheliga, B. M., L. Riggio, and G. Rizzolatti (1994). "Orienting of attention and eye movements." *Experimental Brain Research* 98(3): 507–522.

Sheliga, B. M., L. Riggio, and G. Rizzolatti (1995). "Spatial attention and eye movements." *Experimental Brain Research* 105(2): 261–275.

Shipp, S. (2007). "Structure and function of the cerebral cortex." *Current Biology* 17(12): R443–R449.

Shipp, S. (2016). "Neural elements for predictive coding." *Frontiers in Psychology* 7: 1792.

Shipp, S., R. A. Adams, and K. J. Friston (2013). "Reflections on agranular architecture: predictive coding in the motor cortex." *Trends in Neurosciences* 36(12): 706–716.

Simon, H. A. (1990). "Bounded rationality." In *Utility and Probability*, 15–18. New York: Springer.

Skinner, B. F. (1938). *The Behavior of Organisms: An Experimental Analysis.* New York: Appleton-Century-Crofts.

Smith, R., R. D. Lane, T. Parr, and K. J. Friston (2019). "Neurocomputational mechanisms underlying emotional awareness: insights afforded by deep Active Inference and their potential clinical relevance." *Neuroscience and Biobehavioral Reviews* 107: 473–491.

Smith, R., T. Parr, and K. J. Friston (2019). "Simulating emotions: an Active Inference model of emotional state inference and emotion concept learning." *bioRxiv* 640813.

Solway, A., and M. M. Botvinick (2012). "Goal-directed decision making as probabilistic inference: a computational framework and potential neural correlates." *Psychological Review* 119(1): 120–154.

Sterling, P. (2012). "Allostasis: a model of predictive regulation." *Physiology and Behavior* 106(1): 5–15.

Stewart, N., N. Chater, and G. D. A. Brown (2006). "Decision by sampling." *Cognitive Psychology* 53(1): 1–26.

Stoianov, I., D. Maisto, and G. Pezzulo (2020). "The hippocampal formation as a hierarchical generative model supporting generative replay and continual learning." *bioRxiv* 2020.2001.2016.908889.

Sutton, R. S., and A. G. Barto (1998). *Reinforcement Learning: An Introduction.* Cambridge MA: MIT Press.

Tani, J., and J. White (2020). "Cognitive neurorobotics and self in the shared world, a focused review of ongoing research." *Adaptive Behavior.* 1–20.

Tenenbaum, J. B., T. L. Griffiths, and C. Kemp (2006). "Theory-based Bayesian models of inductive learning and reasoning." *Trends in Cognitive Sciences* 10: 309–318.

Tervo, D. G. R., J. B. Tenenbaum, and S. J. Gershman (2016). "Toward the neural implementation of structure learning." *Current Opinion in Neurobiology* 37: 99–105.

Thomson, A. (2010). "Neocortical layer 6, a review." *Frontiers in Neuroanatomy* 4(13).

Todorov, E. (2004). "Optimality principles in sensorimotor control." *Nature Neuroscience* 7(9): 907–915.

Todorov, E. (2008). "General duality between optimal control and estimation." In *47th IEEE Conference on Decision and Control*, 4286–4292.

Todorov, E. (2009). "Efficient computation of optimal actions." *Proceedings of the National Academy of Sciences USA* 106(28): 11478–11483.

Tolman, E. C. (1948). "Cognitive maps in rats and men." *Psychological Review* 55: 189–208.

Tschantz, A., L. Barca, D. Maisto, C. L. Buckley, A. K. Seth, and G. Pezzulo (2021). "Simulating homeostatic, allostatic and goal-directed forms of interoceptive control using Active Inference." *bioRxiv* 2021.2002.2016.431365.

Tschantz, A., A. K. Seth, and C. L. Buckley (2020). "Learning action-oriented models through active inference." *PLOS Computational Biology* 16(4): e1007805.

Tsvetanov, K. A., R. N. A. Henson, L. K. Tyler, A. Razi, L. Geerligs, T. E. Ham, and J. B. Rowe (2016). "Extrinsic and intrinsic brain network connectivity maintains cognition across the lifespan despite accelerated decay of regional brain activation." *Journal of Neuroscience* 36(11): 3115.

Ueltzhöffer, K. (2018). "Deep Active Inference." *Biological Cybernetics* 112(6): 547–573.

Ungerleider, L. G., and J. V. Haxby (1994). "'What' and 'where' in the human brain." *Current Opinion in Neurobiology* 4(2): 157–165.

van de Laar, T. W., and B. de Vries (2019). "Simulating Active Inference processes by message passing." *Frontiers in Robotics and AI* 6(20).

Veissière, S. P. L., A. Constant, M. J. D. Ramstead, K. J. Friston, and L. J. Kirmayer (2020). "Thinking through other minds: a variational approach to cognition and culture." *Behavioral and Brain Sciences* 43: e90.

Verschure, P., C. M. A. Pennartz, and G. Pezzulo (2014). The why, what, where, when and how of goal-directed choice: neuronal and computational principles. *Philosophical Transactions of the Royal Society of London B: Biological Sciences* 369: 20130483.

Verschure, P. F. M. J. (2012). "Distributed adaptive control: a theory of the mind, brain, body nexus." *Biologically Inspired Cognitive Architectures* 1: 55–72.

Verschure, P. F. M. J., T. Voegtlin, and R. J. Douglas (2003). "Environmentally mediated synergy between perception and behaviour in mobile robots." *Nature* 425(6958): 620–624.

Vincent, P., T. Parr, D. Benrimoh, and K. J. Friston (2019). "With an eye on uncertainty: modelling pupillary responses to environmental volatility." *PLOS Computational Biology* 15(7): e1007126.

Vossel, S., M. Bauer, C. Mathys, R. A. Adams, R. J. Dolan, K. E. Stephan, and K. J. Friston (2014). "Cholinergic stimulation enhances Bayesian belief updating in the deployment of spatial attention." *Journal of Neuroscience* 34(47): 15735.

Wainwright, M. J., and M. I. Jordan (2008). "Graphical models, exponential families, and variational inference." *Foundations and Trends in Machine Learning* 1(1–2): 1–305.

Wald, A. (1947). "An essentially complete class of admissible decision functions." *Annals of Mathematical Statistics* 18(4): 549–555.

Wall, N. R., M. De La Parra, E. M. Callaway, and A. C. Kreitzer (2013). "Differential innervation of direct- and indirect-pathway striatal projection neurons." *Neuron* 79(2): 347–360.

Wesson, D. W., and D. A. Wilson (2011). "Sniffing out the contributions of the olfactory tubercle to the sense of smell: hedonics, sensory integration, and more?" *Neuroscience and Biobehavioral Reviews* 35(3): 655–668.

Wiener, N. (1948). *Cybernetics: or Control and Communication in the Animal and the Machine.* Cambridge, MA: MIT Press.

Winn, J., and C. M. Bishop (2005). "Variational message passing." *Journal of Machine Learning Research* 6(April): 661–694.

Wolpert, D. M., K. Doya, and M. Kawato (2003). "A unifying computational framework for motor control and social interaction." *Philosophical Transactions of the Royal Society of London B: Biological Sciences* 358(1431): 593–602.

Wolpert, D. M., and M. Kawato (1998). "Multiple paired forward and inverse models for motor control." *Neural Networks* 11(7–8): 1317–1329.

Wolpert, D. M., and M. S. Landy (2012). "Motor control is decision-making." *Current Opinion in Neurobiology* 22(6): 996–1003.

Yager, L. M., A. F. Garcia, A. M. Wunsch, and S. M. Ferguson (2015). "The ins and outs of the striatum: role in drug addiction." *Neuroscience* 301: 529–541.

Yamashita, Y., and J. Tani (2008). "Emergence of functional hierarchy in a multiple timescale neural network model: a humanoid robot experiment." *PLOS Computational Biology* 4(11): e1000220.

Yuan, R., and P. Ao (2012). "Beyond Itô versus Stratonovich." *Journal of Statistical Mechanics: Theory and Experiment* 2012(07): P07010.

Yuille, A., and D. Kersten (2006). "Vision as Bayesian inference: analysis by synthesis?" *Probabilistic Models of Cognition* 10(7): 301–308.

Zeki, S., and S. Shipp (1988). "The functional logic of cortical connections." *Nature* 335(6188): 311–317.

Zénon, A., O. Solopchuk, and G. Pezzulo (2019). "An information-theoretic perspective on the costs of cognition." *Neuropsychologia* 123: 5–18.

Zhang, Z., S. Cordeiro Matos, S. Jego, A. Adamantidis, and P. Séguéla (2013). "Norepinephrine drives persistent activity in prefrontal cortex via synergistic α_1 and α_2 adrenoceptors." *PLOS ONE* 8(6): e66122.

Zhou, Y., P. Zeidman, S. Wu, A. Razi, C. Chen, L. Yang, J. Zou, G. Wang, H. Wang, and K. J. Friston (2018). "Altered intrinsic and extrinsic connectivity in schizophrenia." *NeuroImage: Clinical* 17: 704–716.

Index